COUNTDOWN
THE COMPLETE GUIDE TO MODEL ROCKETRY

MICHAEL BANKS

TAB BOOKS Inc.
Blue Ridge Summit, PA 17214

This book is dedicated to model rocketeers everywhere, and especially to the members of the Southwestern Ohio Association of Rocketry (SOAR).

FIRST EDITION

FIRST PRINTING

Printed in the United States of America

Library of Congress Cataloging in Publication Data

Banks, Michael A.
Countdown : the complete guide to model rocketry.

Bibliography: p.
Includes index.
1. Rockets (Aeronautics)—Models. I. Title.
TL844.B36 1985 629.46'52'0228 85-2683
ISBN 0-8306-1991-7 (pbk.)

Front cover photograph courtesy of Centuri Engineering Company.

Contents

Foreword

So you are thinking about flying model rockets? I urge you to do so! They are fun, inexpensive, and safe. I have been flying model rockets for 15 years and have seen thousands of them flown all over the United States. I have seen them flown for the fun of it, in competition for trophy prizes, and on college campuses for scientific research. It makes no difference what your interests are—you can have fun flying model rockets.

Throughout the country there are hundreds of local model rocketeer clubs. These clubs have regular meetings to show films, have sport and competition flies, and to travel to neighboring cities for competition with other model rocket clubs. The clubs pool their resources to build their own launch systems, have their own club flying fields, and have club range equipment such as walkie-talkies and tracking telescopes. I know of some clubs with their own computers, which they use to run flight simulations and check preflight stability.

Model rocketry is an international hobby: there are model rocketeers in Poland, Bulgaria, England, Spain, Canada, Yugoslavia, Australia, and the Soviet Union. If you are really ambitious, you could fly for the United States at the World Model Rocket Championships! These championships are held every other year and are sponsored by the Federation Aeronautique Internationale in Paris, France.

You can also fly model rockets at the United States Model

Rocket Championships. These are held every year and are flown at different locations around the country. This year they might be flown in your own city, where you could become the United States Model Rocket Champion for your age division. The United States Championships are sponsored by the National Association of Rocketry (NAR), headquartered in Elizabeth, Pennsylvania.

If you want to fly model rockets, but feel that you need to improve your skills before competing in the United States, or even the World Championships, you can participate in skill advancement programs. These programs are offered by many of the model rocket manufacturers and by the NAR. In these programs, you proceed at your own pace from one level of difficulty to the next. Each time you reach a new level, you receive a jacket patch, a certificate, and have your name published in a national model rocket magazine.

If you are interested in science and mathematics, model rockets are ideal for research and development. Some college campuses (such as Massachusetts Institute of Technology in Cambridge, Massachusetts, and Ohio State University in Columbus, Ohio) have clubs of students who are doing advanced research with model rockets. Model rockets are lofting sophisticated electronic payloads to sample the atmosphere for pollution, and they are lofting lightweight movie cameras to survey landscapes and study the dynamics of flight.

There are dozens of types of model rockets that you can enjoy flying. For instance, you can fly egglofters, streamer duration birds, gliders, altitude models, and helicopter recovery models. With some practice and experience, you could even fly radio-controlled, model rocket boost-gliders. These gliders are launched piggyback on a model rocket and are then released at the peak of the flight. Using radio control, the gliders are kept in the air for as long as one hour.

You can design and build detailed models that are scaled down versions of actual rockets and missiles, such as the Saturn V moon rocket and the space shuttle. Your scale models can be as detailed and complex in design as you want.

Since 1960, more than 200,000,000 model rockets have been flown throughout the world. Model rocketry is a hobby enjoyed by millions of people of all ages. You, too, can enjoy the hobby. You, too, can be an important part of an active, dynamic sport. I urge you to do so!

J. Patrick Miller, President
National Association of Rocketry

Preface

Model building has many applications. In industry, models are used to design and test potential products; in architecture, models present the aspects of planned buildings; for the military, models can be used to stage mock infantry and armor battles.

Perhaps the most popular application of model building is that of a recreational hobby. No one knows when the first model was constructed, nor what it represented, but the earliest examples of model building known were meant to be enjoyed. (A model motor car, for example, was built in 1679 and patterned after a model steam engine constructed several years earlier.)

Today, models are built by hobbyists for several reasons—as a test of skill, to simulate the workings of full-size vehicles, or simply to be enjoyed on display.

Whatever the reasons, the benefits are the same—the modeler is able to throw out the worries and cares of daily living and immerse himself in a world of his own design, under his own control. In this world, the goals are attainable and the reward is just plain fun and relaxing.

Model rocketeers share in the benefits of model building, of course. The hobby of model rocketry offers an extra bonus, too: the satisfaction of carrying an idea from the initial concept through to its final realization and success. Such activity is true success on the personal level, and there are few experiences to match it!

Acknowledgments

I wish to gratefully acknowledge the following:

For aid in assembling information and materials—Robert L. Cannon, Jeff Flygare, and Mary Roberts, all of Estes Industries, Inc.

For help with photography and illustration—Dick and Barb Buehner, Rich Cardillo, Tom Kolis, and Kevin Wolfe.

For general information, inspiration, and support—Rosa Banks, Tom Hoelle, Doug Pratt, and Bill Brohaugh.

Introduction

Welcome to the exciting hobby and sport of model rocketry! Is the miniature technology of designing, building, and flying model rockets new to you? If so, this book is for you. If you've flown a few models and feel that you've done all there is to do in the hobby, stick around. You'll find quite a few new ideas, as well as some handy references here.

For those of you who are new to model rocketry, some background information may be in order. Model rocketry did not exist until 1957, when G. Harry Stine, an aerospace engineer and writer, was first introduced to the model rocket engine.

At the time, Stine was writing a column on space and rocketry for *Mechanix Illustrated* magazine, and he was concerned with the many accidents connected with home experiments in rocketry. (This amateur experimentation was a natural outgrowth of the interest in space and rocketry generated by the launching of Sputnik by the Soviet Union.)

With this in mind, Stine devoted a column to the safety practices used by real rocket engineers, with some additional words of caution for the accident-prone home builders (or "basement bombers," as they came to be called).

Stine was contacted by Orville H. Carlisle, a shoe store owner and aeronautics hobbyist from Norfolk, Nebraska. Carlisle had developed a safe model rocket engine, drawing on his knowledge of fireworks and pyrotechnics, and was using it to power recoverable rockets in model aeronautics demonstrations.

Stine recognized the potential of the miniature powerplants and eventually began manufacturing model rocket kits and engines. Since then, well over 20 million model rockets have been flown. Thanks to the guidelines set up by Stine, the hobby has an enviable safety record, far better than that of flying model airplanes or even baseball.

Model rocketry also has one of the widest ranges of activity of any hobby. For those who just like to build and fly model rockets as a sport, there are several types of models available. Among these are rockets that simply launch vertically and return via parachute, multistage rockets, and even rockets that sprout wings at their highest altitudes and return to earth as gliders.

For those with a competitive spirit, model rocketry offers endless opportunity in contests on the local, regional, national, and international levels. Categories of competition, which are sanctioned by the National Association of Rocketry, include altitude and duration events, research and development, payload lofting, and scale modeling.

More scientifically oriented rocketeers can likewise find many activities and applications for their interests. For example, it is possible to build and fly aerial photography and telemetry equipment obtained in kit form from the model rocket manufacturers. More advanced rocketeers will find it a challenge to design and build their own rockets and experiments, which can range from cloud chambers and biological sampling packages to rocket-borne television cameras and onboard computers. Computers themselves (the ground-based variety) offer some interesting activities for model rocketeers in the areas of computer-aided design and in performance prediction.

Modelers who work at the craftsman level will find scale modeling an enjoyable activity. The research and exacting construction techniques required to build a flying scale model of a rocket or other aerospace vehicle provide a satisfying challenge.

There are many educational aspects to model rocketry. Model rocketry has been used in education on virtually every level, from grade school to graduate school, and in self-education programs sponsored by model rocket manufacturers and the National Association of Rocketry.

Research and development efforts in science and industry have often involved model rocketry. NASA used specially designed model rocket vehicles to test a laser tracking system, and the

Motorola Corporation at one time used model rockets to test radar tracking systems. Model rockets sometimes replace more expensive full-size rockets in atmospheric sounding experiments, pollution measurement, and high-altitude biological sampling programs. The low cost and versatility of model rockets makes them ideal for these and many other research applications.

Students on the graduate level frequently use model rocketry technology as an adjunct to specialized topics of study. For example, Charles Hall of Ohio State University built a complete computer system that was designed to guide the rocket during its flight and to control the gathering of information through various sensors during flight. A ground-based computer programmed the on-board computer before each flight and then received data stored by the on-board computer after the flight.

On another level, model rockets have been used to test and predict the performance of a variety of aerospace vehicle designs, because they can be designed to precisely duplicate the performance of a prototype. Thus, model rocketry has had applications in real aerospace programs, as well as hobby aerospace activities.

Even though model rocketry was started in the United States, and most rocket manufacturers are located here, the hobby is not limited to this country, or even this continent. Model rockets are flown in Canada, Mexico, all of Europe, Russia, Egypt, and many other countries, and a World Spacemodeling Championship meet is held regularly.

If all of this has given you the idea that model rocketry is a fascinating, fun hobby and sport, you're right. It is the aim of this book to add to your enjoyment of the hobby, and perhaps increase your knowledge in the process. For most rocketeers, fun is the primary consideration, however. If you approach all of your model rocket activities with that in mind, *all* of your launches will be successful.

Fly'em high!

Chapter 1

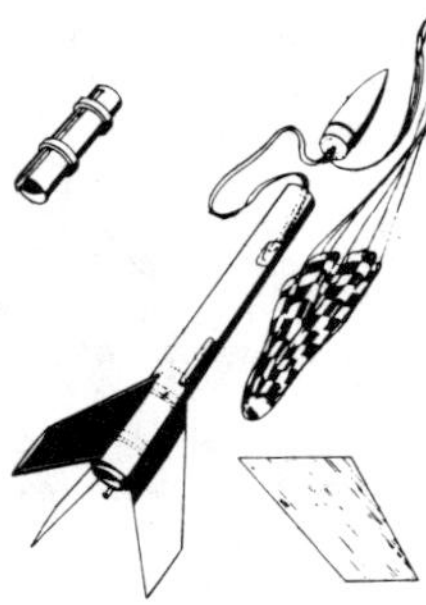

The Basics of Model Rocketry

Any serious discussion of model rocketry should be prefaced with the Model Rocketry Safety Code, developed by the NAR and the Hobby Industry Association of America (HIAA). Flying model rockets in a manner inconsistent with the safety code is definitely dangerous and, in some cases, can be illegal.

The Safety Code is one reason for the excellent safety record enjoyed by model rocketry during the past quarter century. So, in addition to following the manufacturers' instructions in building, flying, and storing model rockets and engines, it is important that you adhere to the Safety Code at all times.

Note the rules of the Model Rocket Safety Code (see box). Please familiarize yourself with it and, again, follow the rules.

Following the rules is simply a matter of applying common sense (although, as Robert A. Heinlein has said, "Common sense is *un*common"). If you stick with these rules, you are almost guaranteed complete safety in your model rocket activities. If you repeatedly violate the rules, you are going to have problems. "Accidents don't just happen," as the saying goes.

You should also note that several of the provisions of the Safety Code are law. Part 101 of the Federal Aviation Administration Regulations (Fig. 1-1) specifies that the weight of a model rocket and the weight of the propellant used in a model rocket engine shall not exceed those limits quoted in the Safety Code. Part 101 also states that model rockets must be made of lightweight, *frangible*

The NAR/HIAA Model Rocketry Safety Code

1. **Construction**—My model rockets will be made of lightweight materials such as paper, wood, plastic, and rubber, without any metal as structural parts.

2. **Engines**—I will use only preloaded factory made engines in the manner recommended by the manufacturer. I will not change in any way nor attempt to reload these engines.

3. **Recovery**—I will always use a recovery system in my model rockets that will return them safely to the ground so that they may be flown again.

4. **Weight Limits**—My model rocket will weigh no more than 453 grams (16 ozs.) at liftoff, and the engines will contain no more than 133 grams (4 ozs.) of propellant.

5. **Stability**—I will check the stability of my model rocket before its first flight, except when launching models of already proven stability.

6. **Launching system**—The system I use to launch my model rockets must be remotely-controlled and electrically operated and will contain a switch that will turn to "off" when released. I will remain at least 15 feet from any rocket that is being launched.

7. **Launch safety**—I will not let anyone approach a model rocket on a launcher until I have made sure that either the safety interlock key has been removed, or the battery has been disconnected from my launcher.

8. **Flying conditions**—I will not launch my model rocket in high winds, near buildings, power lines, tall trees, low flying aircraft, or under any conditions that might be dangerous to people or property.

9. **Launch Area**—My model rockets will always be launched from a cleared area free of any easy to burn materials, and I will only use nonflammable recovery wadding in my rockets.

10. **Jet Deflector**—My launcher will have a jet deflector device to prevent the engine exhaust from hitting the ground directly.

11. **Launch Rod**—To prevent accidental eye injury I will always place the launcher so the end of the rod is above eye level, or cap the end of the rod with my hand or body over the launching rod. When my launcher is not in use I will always store it so that the launch rod is not in an upright position.

12. **Power Lines**—I will never attempt to recover my rocket from a power line or other dangerous place.

13. **Launch Targets & Angles**—I will not launch rockets so their flight path will carry them against targets on the ground, and I will never use an explosive warhead, nor a payload that is intended to be flammable. My launching device will always be pointed within 30 degrees of vertical.

14. **Pre-launch Test**—When conducting research activities with unproven designs or methods, I will, when possible, determine their reliability through prelaunch tests. I will conduct launchings of unproven designs in complete isolation from persons not participating in the actual launching.

(easily bent or broken) materials.

The National Fire Protection Association (NFPA) has established a "Model State Fireworks Law" that specifically excludes model rockets (Fig. 1-2) from the "Fireworks" category. The United States Department of Transportation (DOT) classifies model rocket engines as "Class C Explosives" and "Toy Propellant Devices." (The DOT classification means that model rocket engines are considered safe enough to mail!)

Reprint from FEDERAL AVIATION REGULATIONS PART 101
pertaining to launching of model rockets

Title 14–Aeronautics and Space
Chapter 1–Federal Aviation Agency
Subchapter F–Air Traffic and General Operating Rules [New]

Part 101—Moored Balloons, Kites, Unmanned Rockets, [and Unmanned Free Balloons] [New]

Subpart A—General

§101.1 Applicability.

(a) This Part prescribes rules governing the operation, in the United States, of [the following:]

(1) Any balloon that is moored to the surface of the earth or an object thereon and that has a diameter of more than six feet or a gas capacity of more than 115 cubic feet.

(2) Any kite that weighs more than five pounds and is intended to be flown at the end of a rope or cable.

(3) Any unmanned rocket except—

(i) Aerial firework displays; and

(ii) Model rockets—

(*a*) Using not more than four ounces of propellant;

(*b*) Using a slow-burning propellant;

(*c*) Made of paper, wood, or breakable plastic, containing no substantial metal parts and weighing not more than 16 ounces, including the propellant; and

(*d*) Operated in a manner that does not create a hazard to persons, property, or other aircraft.

[(4) Any unmanned free balloon that—

[(i) Carries a payload package that weighs more than four pounds and has a weight/size ratio of more than three ounces per square inch on any surface of the package, determined by dividing the total weight in ounces of the payload package by the area in square inches of its smallest surface;

[(ii) Carries a payload package that weighs more than six pounds;

[(iii) Carries a payload, of two or more packages, that weighs more than 12 pounds; or

[(iv) Uses a rope or other device for suspension of the payload that requires an impact force of more than 50 pounds to separate the suspended payload from the balloon.]

(b) For the purposes of this Part, a "gyroglider" attached to a vehicle on the surface of the earth is considered to be a kite.

§101.3 Waivers.

No person may conduct operations that require a deviation from this Part except under a certificate of waiver issued by the Administrator.

§101.5 Operations in prohibited or restricted areas.

No person may operate a moored balloon, kite, unmanned rocket, [or unmanned free balloon] in a prohibited or restricted area unless he has permission from the using or controlling agency, as appropriate.

Fig. 1-1. Excerpt from FAA Part 101, pertaining to model rockets. Courtesy Centuri Engineering Co.

RESTRICTIONS

Despite the excellent safety record of the hobby, the recommended Safety Code, and the "official" classification of model rockets and model rocket engines, you will find some local restrictions—usually state laws—affecting model rocketry. The following states either require permits for the purchase of model rocket engines, or have other restrictions pertaining to the model rocketeer's age or to launch sites: California, Massachusetts, New

Fig. 1-2. Simple model rocket.

Jersey, Oregon, and Rhode Island. Check with your state Fire Marshal's Office or with model rocket manufacturers for details.

Even with restrictions, model rocketry is fun. You will find that any problems with local authorities can usually be eliminated by making sure that you have permission to use a launch site, by following the Model Rocket Safety Code, and by observing any other local requirements. You will also find it a plus to belong to the NAR and carry NAR insurance.

FOUR KINDS OF ROCKETRY

One reason that the restrictions just discussed exist is the fact that many people confuse model rocketry with amateur rocketry and "basement bomber" activities. So that you will know the difference, I will define model rocketry in comparison with other types of rocketry.

The crudest (and most dangerous) form of rocketry is that practiced by the *basement bomber*. Typically, the basement bomber throws together an explosive mixture, stuffs it into a pipe, attaches some metal fins, and promptly blows himself to bits when he ignites the mess. Some basement bombers are less sophisticated, using match heads instead of a chemical mixture, but the results are always the same. That is, the "rocket" will be monumental failure, and the "rocketeer" faces a one in seven chance of being maimed or killed.

Statistics—hard facts—back these statements up. And, contrary to what you might think, the basement bomber is not always a youngster; adults (even science teachers working on a class project) have been killed while experimenting with home-brewed rocket fuel.

In essence, being a basement bomber is suicide. It is also illegal, and is just what model rocketry was designed to prevent and replace.

Amateur rocketry is something quite apart from the activities of the basement bomber. Those involved in amateur rocketry follow even stricter rules than model rocketeers in the design, construction, and flying of their rockets. They also mix their own fuels, but they are experts at it and are aware of the risks.

Amateur rocketeers are subject to very stringent government regulation and are usually engaged in serious scientific research. Amateur rocketry is a very expensive proposition in terms of the machinery, equipment, training, and materials necessary. It is almost always a group activity, and there are very few amateur rocketry organizations in this country. (You may be interested to know that Robert H. Goddard was America's first amateur rocketeer.)

Postscript: amateur rocketeers experience a high failure rate!

Professional rocketry is carried out by governments and aerospace manufacturers such as North American Rockwell, Thiokol Chemical Corporation, and Boeing Aerospace. You'll find professional rocketeers at Cape Canaveral, the European Space Agency's French Guiana launch site, and at the Baikonur Cosmodrome in Russia, among very few other places (Fig. 1-3). Professional rocketry requires years of training and billions of dollars, and is always concerned with scientific research and military applications.

All of this should give you a feel for what *model rocketry* is, and is not. In model rocketry, you will find almost all of the benefits of professional rocketry, a few restrictions, and a very low cost. What you will *not* find in model rocketry are any of the dangers of other forms of rocketry, nor the expense. For the most part, you will be using ready-made rockets. If you build your own rockets, you will be using tested, workable designs. The fuel you use will also be tested, commercially prepared solid fuel. What more can you ask for?

BASIC COMPONENTS OF A MODEL ROCKET

Let's take a closer look at the specific components of a model rocket. By definition, a model rocket is a miniature, ultra-light-

Fig. 1-3. Space shuttle launch, April 12, 1981. Courtesy NASA.

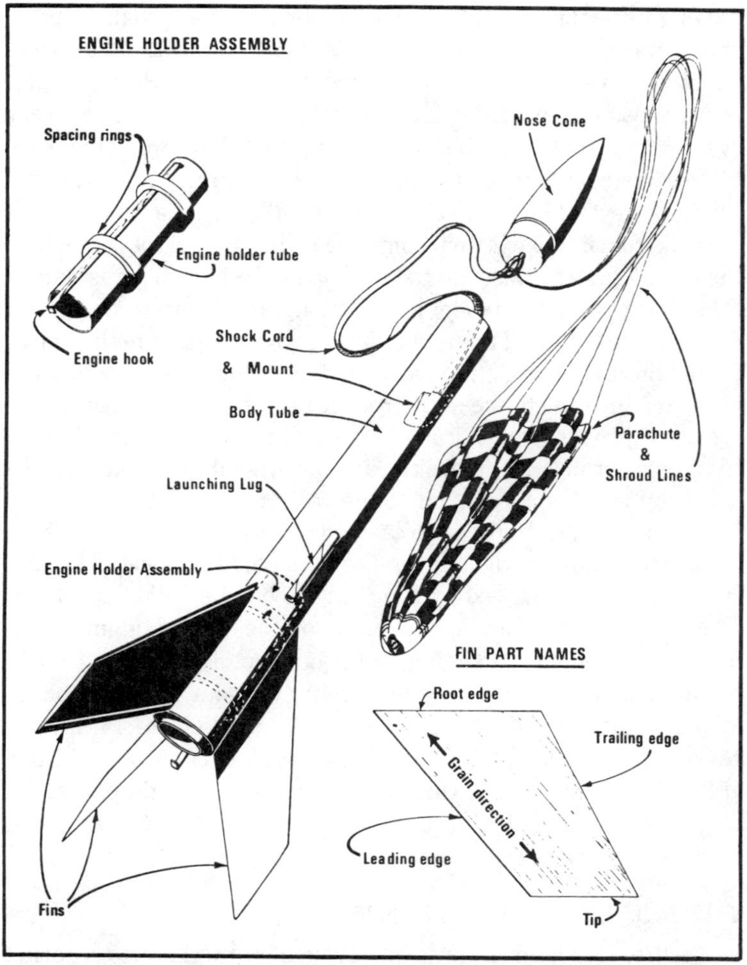

Fig. 1-4. Basic components of a model rocket. Courtesy Estes Industries.

weight solid fuel rocket. Figure 1-4 illustrates the basic components of a typical model rocket. (Model rockets come in many different shapes and sizes.)

The primary element of a model rocket is the *model rocket engine*. It is a safe, nonreusable device. A *body tube*, or *air frame*, is the basic structural component of a model rocket. The body tube is made of strong lightweight cardboard, specially treated to be flame resistant. Every other part of a model rocket is either attached to or depends upon the body tube.

Inside the body tube, there is an *engine mount*, designed to hold the engine securely in place. The *recovery system* (usually a parachute attached to the body tube via its *shroud lines* and a *shock cord*) guarantees a safe descent for the model rocket, as well as preventing the loss of the model. Between the engine and the recovery system, there is flameproof *recovery wadding*. This protects the recovery system from the hot exhaust gases of the engine that activate the system and serve as a "piston" during activation.

The remaining external components of a model rocket include the *launch lug*, the *fins*, and the *nose cone*. The launch lug is a small tube that guides the model rocket along a rod during the first 3 feet of its flight, until the rocket has reached a speed high enough for the fins to become effective in guidance. The fins, through their aerodynamic effects, keep the rocket in a stable flight path. Fins are usually made from balsa or plastic. The nose cone, also formed from balsa or plastic, guides the airflow around the rocket (reducing drag) and helps to stabilize the rocket.

Figure 1-5 illustrates a typical model rocket flight profile. There are seven separate stages in the flight of a model rocket. The first step is *ignition*, followed by *liftoff* as the engine builds up thrust. The time of thrust varies, depending on the type of engine used, but *burnout* (when the engine fuel is completely used up) usually occurs after two or three seconds. After burnout, the *delay* phase of the flight begins and continues as the rocket coasts upward several hundred (or as much as a thousand) feet beyond its altitude at burnout. At *apogee*, or the highest point of the flight curve, the recovery system is deployed. Following this is the *descent phase*, with eventual *recovery*.

THE MODEL ROCKET ENGINE

All of this is made possible, of course, by the model rocket engine. Model rocket engines are miniature power plants with several functions, as illustrated in Fig. 1-6. Figure 1-7 shows several types of available model rocket engines. The components of the model rocket engine are the *casing*, which is tightly wound paper or fiberglass; a clay or graphite *nozzle*; the *propellant*, which provides the thrust; the *delay element*, which is a slow-burning, pretimed charge ignited by the propellant at burnout; the *ejection charge*, which is ignited by the delay element and produces pressurized gases to activate the recovery system; and the *cap*, a clay plug above the ejection charge.

Figure 1-8 provides a graphic explanation of a model rocket engine's functions as they relate to the flight of a model rocket.

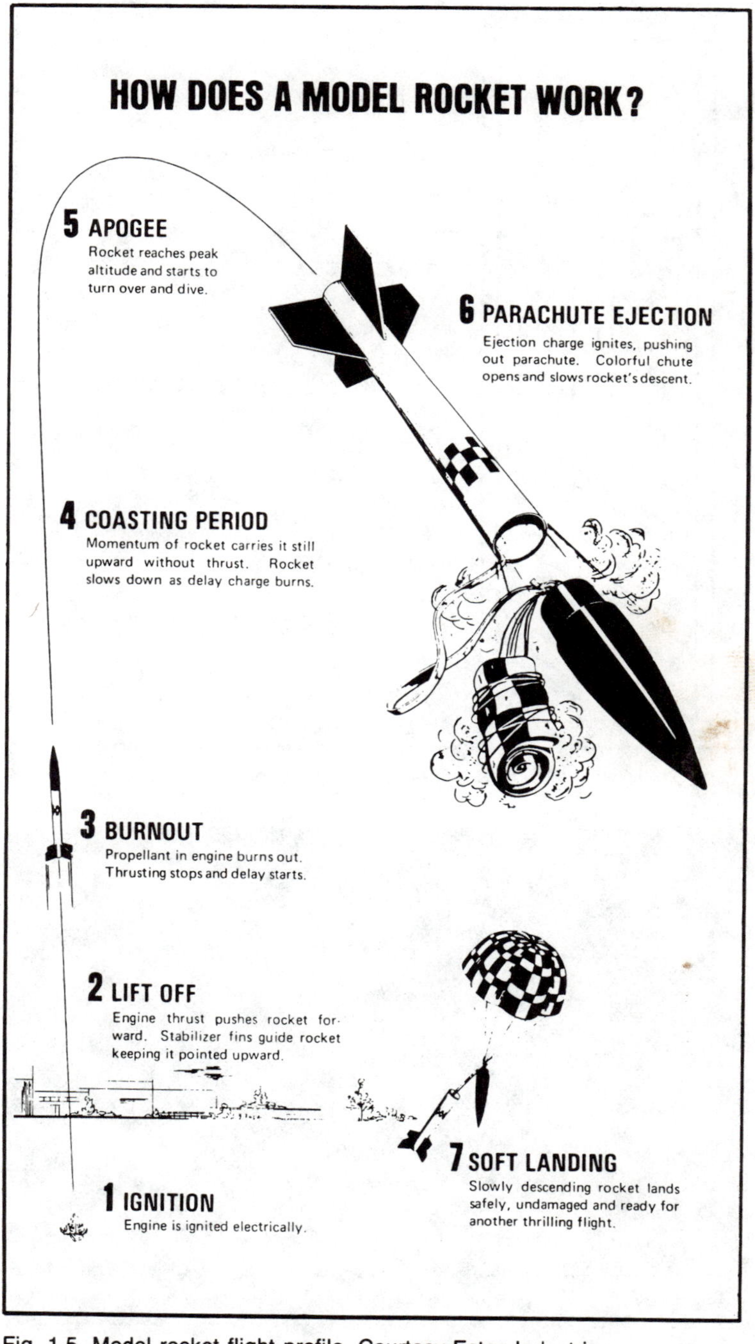

Fig. 1-5. Model rocket flight profile. Courtesy Estes Industries.

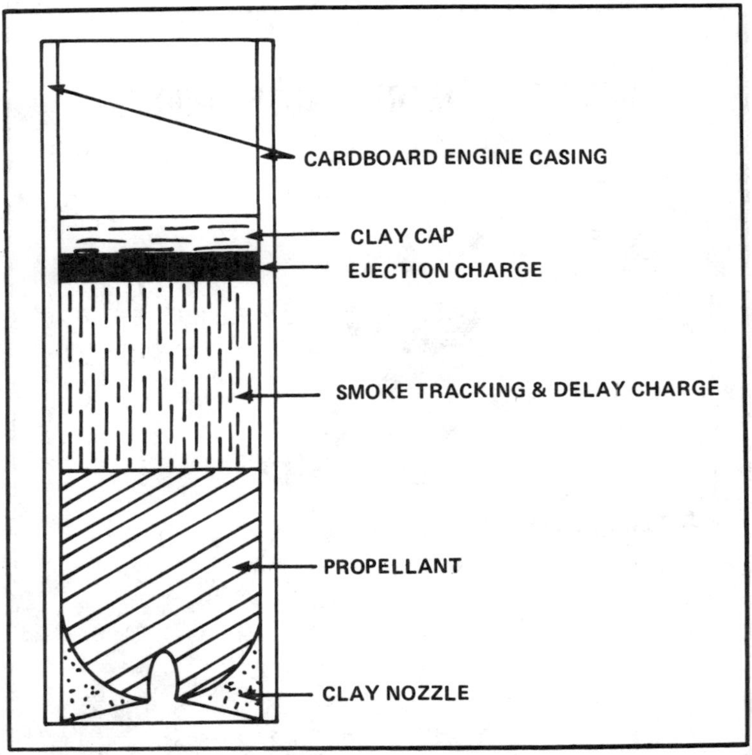

Fig. 1-6. Cutaway view of a standard model rocket engine.

Fig. 1-7. Typical model rocket engines.

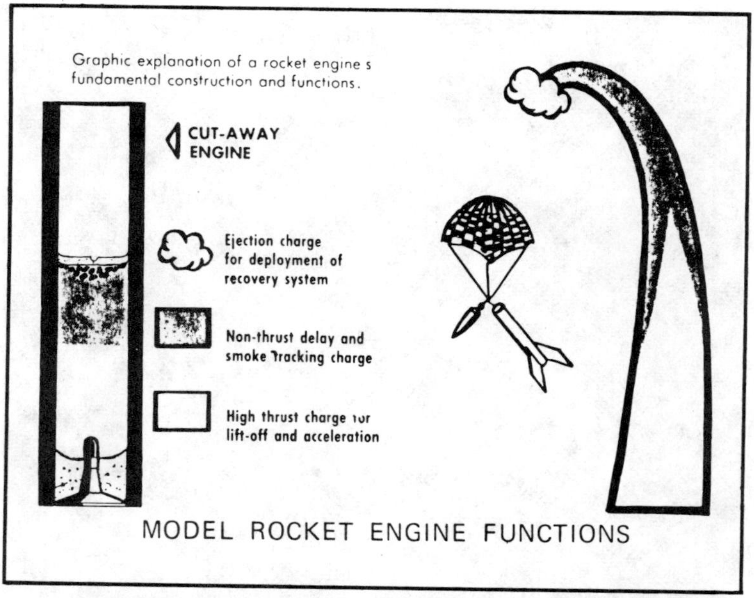

Fig. 1-8. Model rocket engine functions. Designed for specific launch and flight functions, rocket engines are not re-usable. Courtesy Centuri Engineering Co.

All model rocket engines carry an alphanumeric code (Fig. 1-9) that gives the *total impulse* (total power) of the engine in *newton-seconds*, the average thrust of the engine in newton-seconds, and the delay time between burnout and activation of the ejection charge. (A newton-second is defined as the force that produces an acceleration of one meter per second on a mass of one kilogram.)

The first character in the code will be a letter, A through F. The letter indicates the total impulse and may be "translated" into newton-seconds using the data in Table 1-1.

As you can see, each engine has twice the total impulse of its predecessor. (There are also available 1/4A and 1/2A engines, of .625 and 1.25 newton-seconds total impulse, respectively.) It may help you to know that 4.12 newton-seconds is equal to 1 pound-second. Newton-second and *pound-second* are force/time measurements, indicating the amount of energy expended in one second.

If you examine the code on a particular engine (Fig. 1-10), you will find that (through some simple math) you can obtain more information about the engine than is given. A "D12-5" engine, for example, tells you on the label that it has a total impulse of 20 newton-seconds, an average impulse of 12 newton-seconds, and a delay time of 5 seconds.

Fig. 1-9. Model rocket engines come in clearly marked packages. Courtesy Estes Industries.

If you divide the total impulse (20 n-sec.) by the average impulse (12 n-sec.), you will know the thrust time of the engine, because the average impulse is measured over the entire time that the propellant is burning. Thus, a D12 engine has a thrust time of

Table 1-1. Letter Values in Model Rocket Engine Coding.

LETTER	TOTAL IMPULSE
A	2.50 newton-seconds
B	5.00 newton-seconds
C	10.00 newton-seconds
D	20.00 newton-seconds
E	40.00 newton-seconds
F	80.00 newton-seconds

Fig. 1-10. Cutaway view of a high-power engine. Courtesy Centuri Engineering Co.

1.7 seconds, if you round off the decimals.

You already know the delay time (5 seconds), and if you add the thrust duration (1.7 seconds) to this, you know that the rocket should reach apogee (the highest part of its flight) approximately 6.7 seconds after liftoff, to insure optimum recovery system performance. If the rocket is still coasting upward when the ejection charge (and thus, the recovery system) is activated, the rocket could, for example, travel right through the parachute! If the rocket reaches apogee long before ejection, it could be falling at a rate fast enough to rip the recovery system to pieces when the ejection charge is activated.

RECOVERY SYSTEMS

Parachutes are not the only type of recovery system, of course. There are five basic recovery methods. Figure 1-11 illustrates these methods.

The *parachute* system is perhaps the most frequently used. The usual system consists of a parachute made of plastic or nylon, the parachute's shroud lines, and the shock cord. In a typical rocket, the parachute is attached by its six or eight shroud lines to the base of the nose cone, and the shock cord, which is made from an elastic material, connects the nose cone to the body tube. When the hot gases of the ejection charge begin expanding in the body tube, the flameproof recovery wadding is pushed forward, against the parachute, in a piston effect. This action pushes the nose cone out of the body tube, and it is followed by the parachute. The force of the ejection, along with the airflow caused by the movement of the rocket, forces the parachute open, and the shock cord stretches to take up the shock of the parachute's opening.

The *streamer* system of recovery is similar in construction to the parachute system, but instead of a parachute, a streamer is used. The streamer is usually made from flameproof crepe paper, but may sometimes consist of plastic coated on one side with mylar to enhance visibility.

A rocket with a streamer recovery system will descend much faster than one with a parachute. This is useful if you are flying in a small field, because the rocket will have less time to drift out of the recovery area. If you are flying in windy conditions, a streamer is mandatory. (You should never fly in winds over 15 mph, in any event.) If you plan to use a streamer, you should take into account the fact that a heavy rocket could be damaged by the faster descent rate permitted by a streamer. You can, of course, increase the drag of a streamer (and thereby slow the descent) by making it longer or wider, or by using two or even three streamers.

Glide recovery is just what the name implies. A rocket using glide recovery goes up like a rocket, and returns like an airplane (similar to the space shuttle in concept).

One type of rocket using glide recovery, known as the *boost-glider*, looks much like a regular hand-launched glider, with one exception. The boost-glider has a long boom on its nose, on which the engine mount is fixed. The entended boom moves the boost-glider's center of gravity, thus making it stable during vertical flight. At ejection, the engine either falls away (to be recovered by

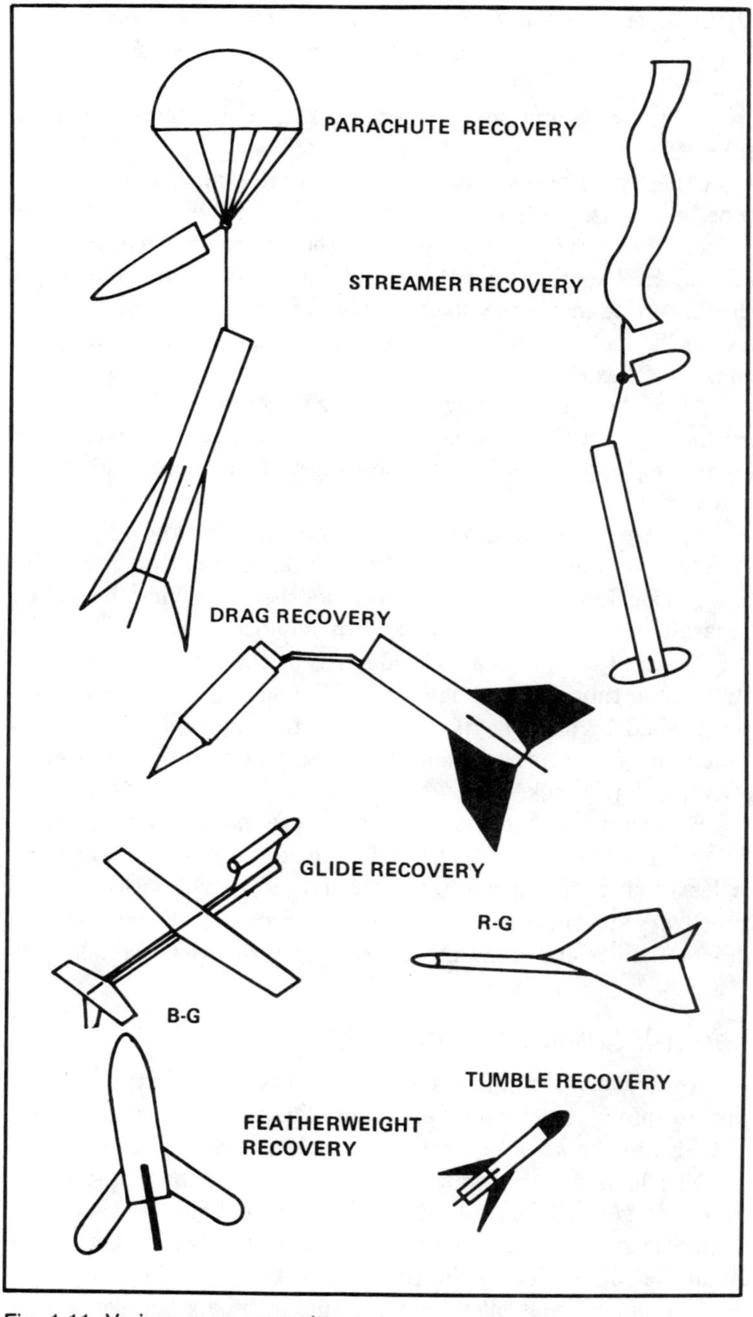

Fig. 1-11. Various recovery systems.

streamer or parachute), or is moved back toward the glider wing. In either case, the center of gravity is moved rearward, resulting in a stable glider.

Technically, any boost-glider that does not eject or separate from its engine is a *rocket-glider*.

Other types of boost gliders may sprout wings from the body tube in a scissors effect, or they may ride on the side of a large rocket (gliders riding on the side of a rocket are known as *parasite gliders*). Still another type of boost-glider involves a flexible glider that is carried inside the body tube of the host rocket, and pushed out, or "launched," at ejection. This form of glider is called a *flex-wing*, or "flexie."

All of these gliders have one thing in common: the rudder or another control surface is set so that the glider returns to earth in a spiraling path. This insures the return of the glider within the recovery area.

The *drag* method of recovery requires that the rocket's shape be altered at ejection to produce a high amount of drag, or resistance to air flow. This is usually accomplished by having the rocket separate into two parts connected by a shock cord.

Tumble recovery is a method whereby the rocket is made unstable, and thus tumbles as it falls. The instability is usually accomplished by shifting the position of the engine (and thus the center of gravity) at ejection. The tumbling causes a high drag effect, and the rocket falls to earth very slowly.

Both drag and tumble recovery are best suited to very lightweight model rockets. Some few rockets are so light (3 grams or less) that they can fall to the ground safely without the aid of a recovery system. Although this is not a recovery system as such, this method is known as the *featherweight* recovery method.

GROUND SUPPORT EQUIPMENT

Ground support equipment (GSE) consists of everything necessary to safely launch a model rocket. The two basic components of GSE are the *launcher* and the *launch controller* (Fig. 1-12).

The launcher (or launch pad) consists of a base (usually a tripod), a *blast deflector* (to deflect the hot exhaust gases of the engine away from the ground), and a *launch rod* or *launch rail* to guide the rocket during the first few feet of its flight.

The launch controller's components include a battery, a continuity light, a switch, a safety interlock key, at least 15 feet of wire,

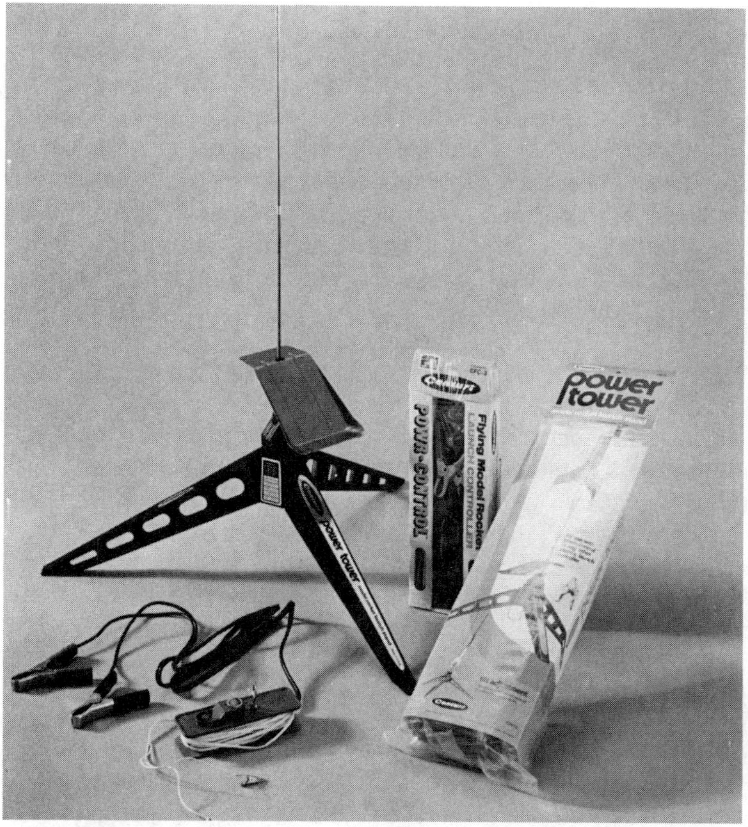

Fig. 1-12. Versatile tripod launch stand and launch controller. Courtesy Centuri Industries.

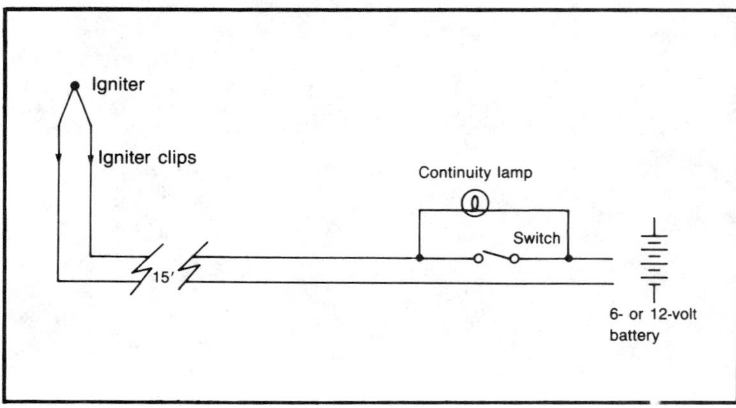

Fig. 1-13. Launch controller schematic.

and micro-clips for connecting the wires to the engine igniter. Figure 1-13 shows the schematic diagram of a typical launch controller.

In use, the launch controller is first connected to a battery (or, the batteries are placed inside the controller, depending on what kind of controller you have), and the micro-clips are connected to the igniter. Then, the safety interlock key is inserted, which "arms" the launcher (Fig. 1-14). At this time, the continuity light will come on, indicating that there is a continuous circuit through the controller and the igniter. (If the light does not come on, you have a problem with a poor connection at the battery or igniter.)

There is some current flowing through the igniter at this stage, but, because of the fact that the lamp is on, the voltage is not high enough to cause the igniter to heat up and ignite the engine. As soon as the launch switch is pressed, however, the lamp is cut out of the circuit, and the full voltage of the battery is applied across the igniter.

This results in the igniter heating up because it is made from *nichrome wire*, a high-resistance wire that becomes hot very quickly. This in turn causes a blob of pyrotechnic material at the end of the

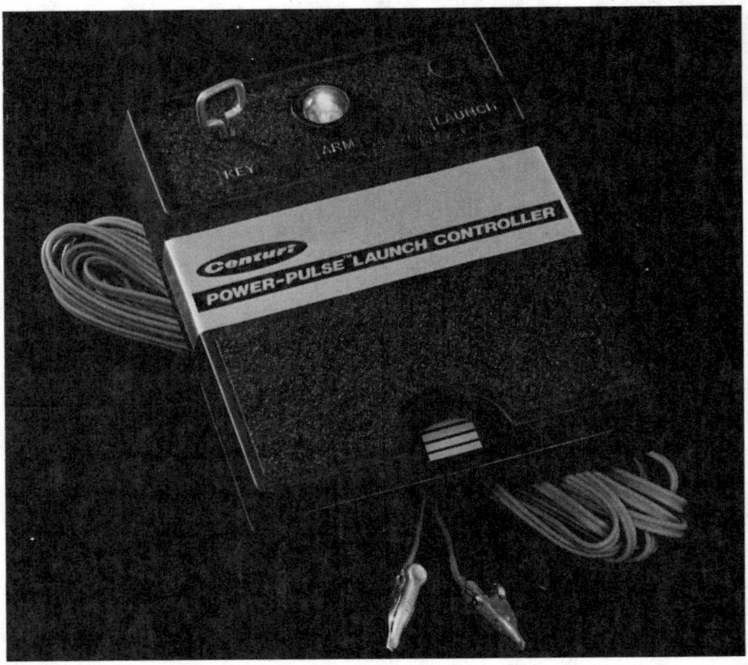

Fig. 1-14. Launch controllers feature a safety key and an "arm" button to prevent accidental launchings. Courtesy Centuri Industries.

Fig. 1-15. Typical engine igniters.

igniter to flare up and this ignites the engine (Fig. 1-15).

Note: You should only use this method—or one similar—to ignite model rocket engines. Fuses are unreliable, dangerous, and sometimes illegal!)

Complete systems, including all components of the launcher and the launch controller, are available from manufacturers, or you may build your own GSE, using the plans on Fig. 1-16.

GETTING STARTED

The easiest way to get started in model rocketry is to purchase one of the many starter kits (Fig. 1-17) offered by the manufacturers. Several are shown in the accompanying photographs. As you can see, there are a variety of launchers and controllers, differing in the placement of the batteries and the appearance of the launcher and controller housing. It is up to you to choose the one that best suits your needs.

Each starter kit includes a model rocket, two or three engines, igniters, recovery wadding, and a manual (Fig. 1-18). Usually, the model rocket is one of the simple kits (Fig. 1-19), although some starter sets are based on more advanced rockets (Fig. 1-20). In either case, you will find that the cost of the starter kit is far less than the total cost of its individual components.

Starter kits such as these (or the individual components), as well as other model rocket kits and supplies may be purchased directly from the manufacturers listed in Appendix A, or from your local hobby shop or department store (Fig. 1-21).

Again, it is possible for you to make your own GSE, in accord-

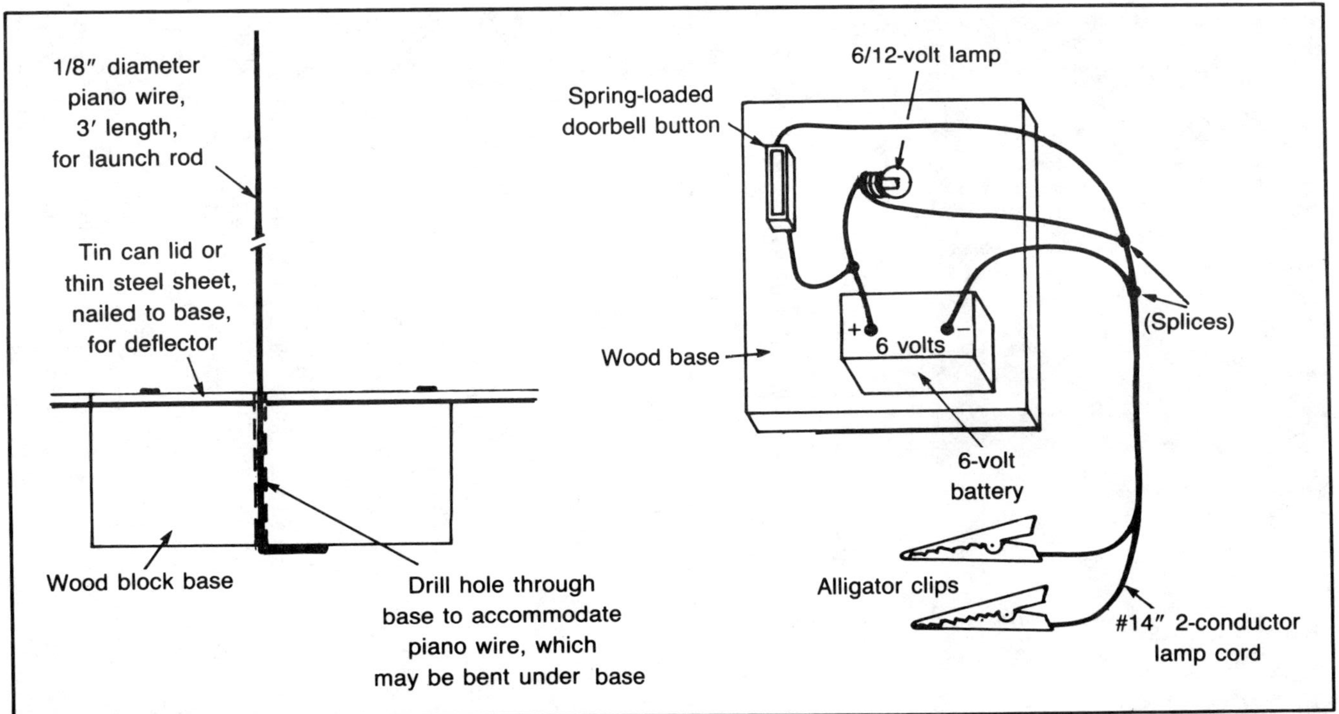

Fig. 1-16. A homemade launch pad and controller.

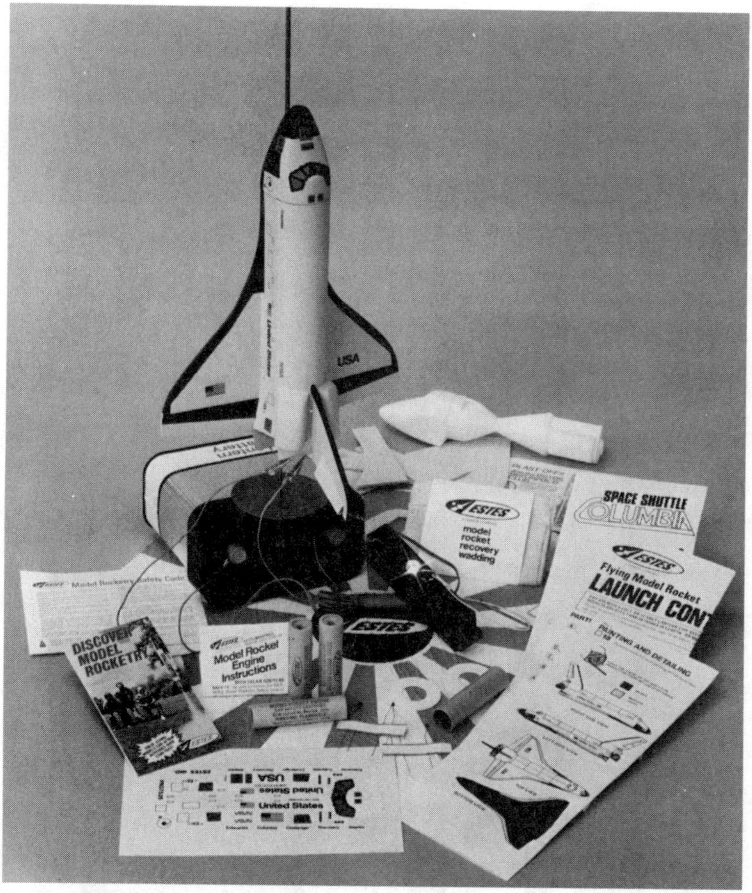

Fig. 1-17. A model rocket starter kit that features the NASA space shuttle, literature, decals, and engines. Courtesy Estes Industries.

ance with the accompanying plans, but you will find it far easier to build model rockets using the kits and parts supplied by the manufacturers, because these are designed for the purpose (Fig. 1-22). Do not, in any case, attempt to load or reload your own engines! This is definitely a dangerous and illegal activity and is unreliable in terms of the performance you will get from the engines (assuming, that is, the entire enterprise doesn't blow up in your face!). Model rocket manufacturers have invested hundreds of thousands of dollars and years of experience in facilities for manufacturing model rocket engines. So why not leave it to the pros?

SELECTED MODEL ROCKET KITS

The model rocket kits described in this section present a cross section of various types of kits (Table 1-2). The kits were chosen on the basis of how well they represent their respective categories and how likely they are to be available in the years to come.

The purpose of the coverage here is not to recommend one kit or line of kits over another, but to provide the beginner with some guidelines for selecting kits to build.

Remember that this is by no means a complete listing of available kits, but a representative sampling. You will be able to find many kits similar to those reviewed here in manufacturers' catalogs.

Beginner Kits

Alpha. Manufactured by Estes Industries, this model rocket is just what the name implies—and excellent first kit (Fig. 1-23).

Fig. 1-18. Another complete starter outfit from Centuri Engineering.

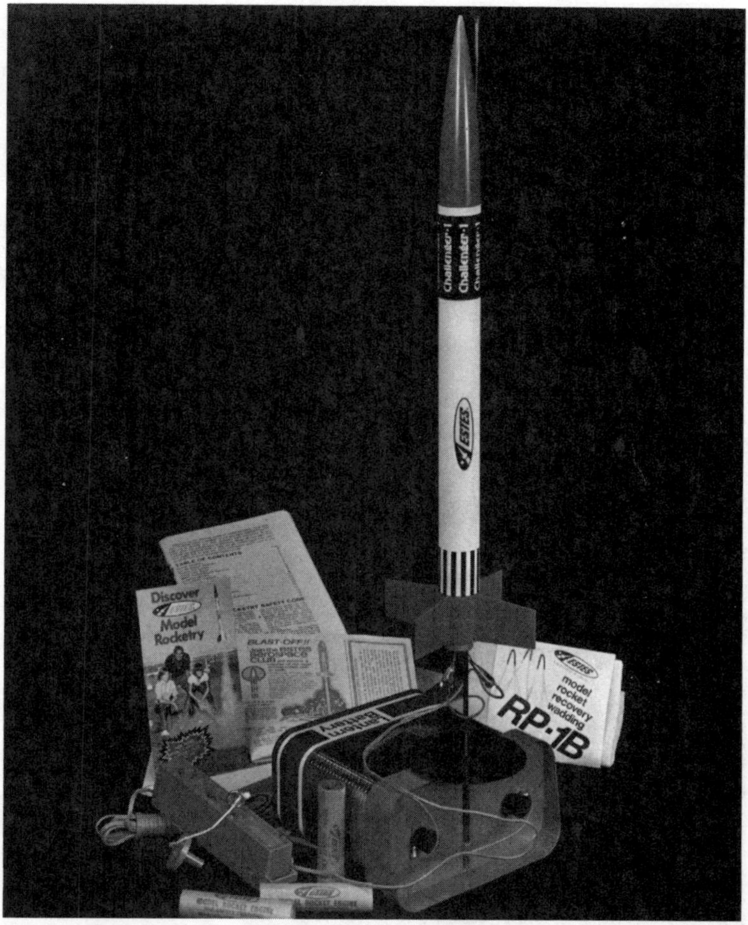

Fig. 1-19. A simple, but complete starter kit is the best way to begin model rocketry. Courtesy Estes Industries.

Requires sanding and painting. Uses A, B, or C engines.

Alpha III. This is an improved version of the Estes Alpha (Fig. 1-24), featuring the simplicity of design and stability of the original combined with a plastic nose cone and plastic fin unit for ease of assembly. Uses A, B, or C engines.

Challenger II. Usable as a carrier for the Astrocam aerial camera, this is also an excellent sport flyer. Offered by Estes, the Challenger II (Fig. 1-25) is a good place to start high-power rocketry. This kit features a plastic nose cone and plastic fin unit and uses the D12-7 engine.

Firebird. A medium size (16.3-inch long) rocket from Centuri Engineering, with plastic nose cone. Uses A, B, or C engines.

Maxi-Alpha III. A "maxi" version of Estes Industries' Alpha III, this rocket is an impressive 33.25 inches in length and is easy to build, using a plastic nose cone and press-fit plastic fin unit. D-powered (Fig. 1-26).

Micro. A product of Flight Systems Inc. (FSI), the Micro is easily convertible from A-type to more powerful engines and is capable of altitudes up to 3000 feet. Can be assembled in 15 minutes.

Phoenix Bird. Centuri's Phoenix Bird is a medium size model (26.3 inches) with a mid-fuselage breakdown for deployment of the

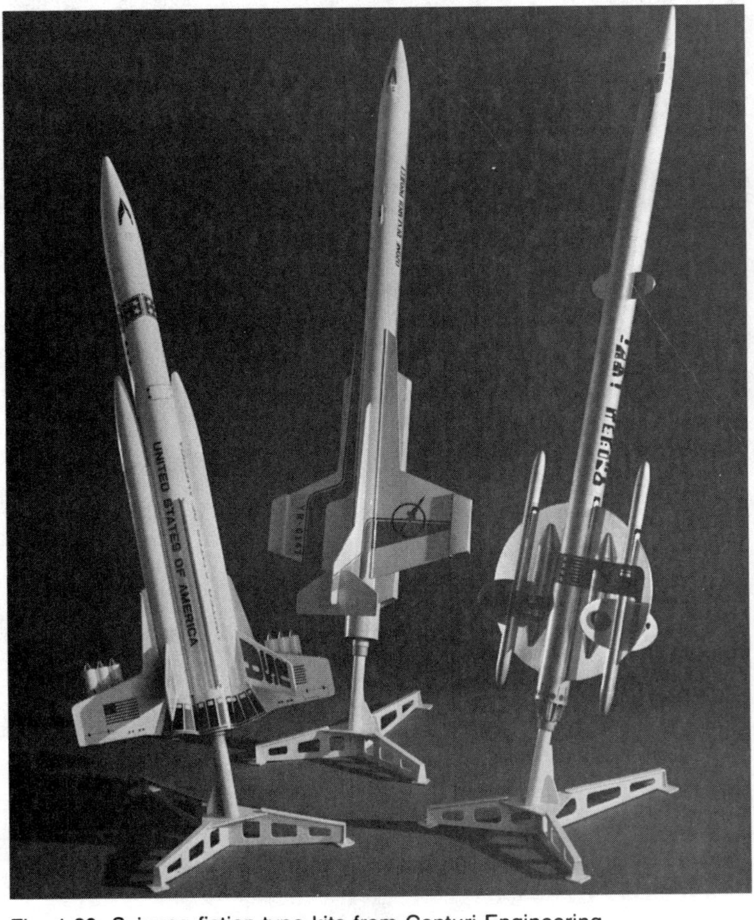

Fig. 1-20. Science fiction-type kits from Centuri Engineering.

Fig. 1-21. Model rockets come in a wide range of sizes. Courtesy Centuri Engineering Co.

recovery system. This is an easy to assemble kit, with plastic nose cone and fin unit, stick-on decals, and precolored body tube. B- or C-powered.

Viking. Another excellent first kit, the Viking (Fig. 1-27), from Centuri, features plastic nose cone, fiber fins, a choice of up to 12 designs, and precolored parts. A, B, or C engines.

Intermediate Kits

Andromeda. This is one of the Estes Industries' more attractive science fiction kits. Over 43 inches long, this starship is B- or C-powered.

ASP. An excellent introduction to scale modeling (Fig. 1-28), Estes' ASP is based on an atmospheric sounding rocket, and is mini-engine (1/2A, 1/2A, or A) powered.

Beta. Estes' Beta is recommended as your first multistage kit. This rocket can be flown with mini-engines (1/4A, 1/2A, or A) and is very stable. May be flown as a single-stage or a two-stage rocket.

Space Shuttle Columbia. A truly unique model rocket, the Columbia is a semiscale model (Fig. 1-29). The bulk of the kit is pre-formed, using *skinned foam* (a vinyl "skin" covering a foam core). Other parts of the kit are made from plastic, which facilitates

Fig. 1-22. A homemade "miniature" model rocket. Courtesy J. Kevin Wolfe.

Table 1-2. Model Rocket Kit Selection Chart.

Range	Beginner								Intermediate							Advanced/ Specialized					Scale									
KIT NAME	Alpha	Alpha III	Challenger II	Firebird	Maxi-alpha III	Micro	Phoenix Bird	Viking	Andromeda	ASP	Beta	Shuttle Columbia	Gyroc	Satellite Killer	Sprint	Astrocam	Echo I	Mars Lander	Maverick	Viking I	Black Brant II	Black Brant III	"Cruise Missile"	LTV Scout	Jayhawk	Mercury Redstone	Phoenix	Saturn V	Space Shuttle	SR-71 Blackbird
TYPE	S	S	S	S	S	S	S	S	S	SC	S/C	S	S	S	C	R	R	S	R	S	SC	SC	SC	SC	SC	SC	SC	SC	SC	SC
SKILL LEVEL	1	1	2	1	2	1	1	1	3	2	2	2	3	3	3	2	3	5	4	3	4	2	4	3	5	5	3	5	4	3
ENGINES USED	A, B C	A, B C	D	A, B C	D	A, B D, E	B, C	A, B C	B, C	M	M	A, B C	A, B C	C	A, B C	B, C	A C, D	B C	E F	A, B C, D	F	A, B C	A, B C	C	D	C	D	C D	C	C D
RECOVERY METHOD	P	P	P	P	P	P	P	S	P	P	P	P	H	P	S	P	S	P	P S	P	P	P	P	P	P	P	P	P	P G	P
PLASTIC	NO	YES	YES	YES	YES	NO	YES	YES	NO	NO	NO	YES	NO	YES	NO	YES	NO	YES	NO	NO	YES	NO	YES	YES	NO	YES	YES	YES	YES	YES
DECALS	YES	YES	YES	YES	YES	YES	YES	YES	YES	YES	YES	YES	YES	YES	YES	YES	YES	YES	YES	YES	YES	YES	YES	YES	YES	YES	YES	YES	YES	YES
MANU-FACTURER	E	E	E	C	E	F	C	C	E	E	E	E	E	C	E	E	F	E	F	F	F	E	E	E	C	C	E	E	E	C

(NOTE: Skill Levels are rated to 1 to 5, with 1 the simplest and 5 the most difficult.)
TYPES: *S*-Sport Flyer, *C*-Competition, *SC*-Scale Model, *R*-Research and/or experimental.
RECOVERY METHODS: *P*-Parachute, *S*-Streamer, *H*-Helicopter, *G*-Glide.
MANUFACTURERS: *E*-Estes Industries, Inc., *F*-Flight Systems, Inc., *C*-Centuri

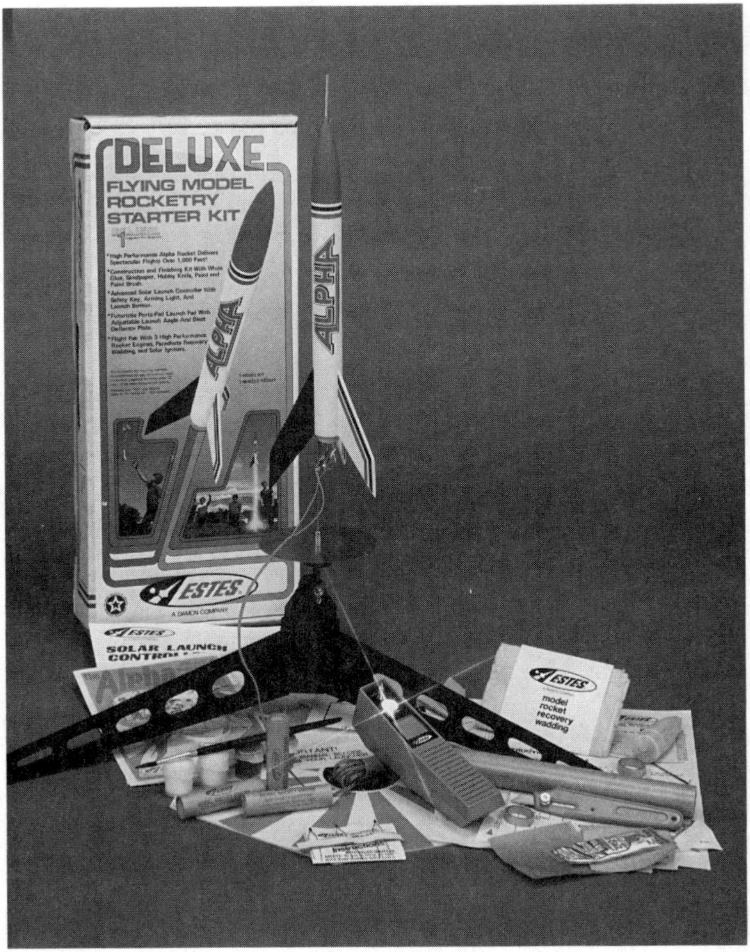

Fig. 1-23. The Estes Alpha starter kit. The Alpha is the most basic of model rockets.

assembly. Some experience in building model rockets is required. 10.7 inches in length, the Columbia is A-, B-, or C-powered. Manufactured by Estes.

Gyroc. An old-time Estes kit, the Gyroc demonstrates helicopter recovery, deploying small fin tips at apogee to cause spinning effect. A-, B-, or C-powered.

Satellite Killer. Manufactured by Centuri, this model rocket resembles a satellite with solar panels deployed. Very impressive flight and recovery (Fig. 1-30). C-powered.

Sprint. This competition model is offered by Flight Systems

Inc. and features a "boat-tail" for reduced drag. The Sprint flies with a large range of low- to high-power engines. Fifteen inches long.

Advanced and Specialized Kits

Astrocam. Estes' Astrocam (Fig. 1-31) is intended for those with some model rocketry experience. The kit includes the camera, which must be assembled, and the booster rocket Delta II. The

Fig. 1-24. The Alpha III, an improved version of the Alpha from Estes, sporting plastic nose cone and fins.

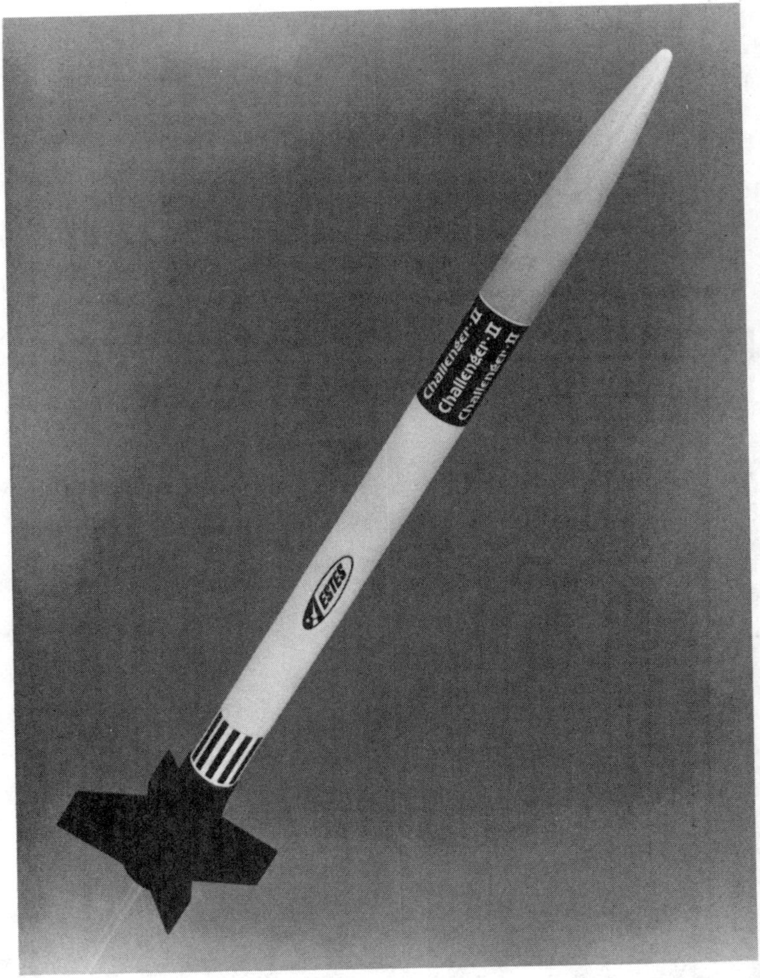

Fig. 1-25. The Challenger II, manufactured by Estes as a high- altitude Astrocam launch vehicle. It is D-powered and has plastic nose cone and fin unit for ease of assembly.

Delta II requires no painting and has a plastic fin unit. B- or C-powered.

Echo I. From FSI, the Echo I is an advanced two-staged model rocket that is capable of flights to extreme altitudes. A-, B-, C-, or D-powered.

Mars Lander. Another Estes favorite, the Mars Lander (Fig. 1-32) features parachute recovery and a landing cushioned by four landing legs that are deployed at apogee. Impressive in flight and on display, the kit requires quite a bit of experience to build. B- or C-powered.

Maverick. Another FSI offering, the Maverick is designed to carry payloads to high altitudes. Twenty-nine inches in length, the Maverick uses D and other high-power engines.

Viking I. FSI's Viking I is the first of a series of kits featuring advanced design features, including tubular stabilizers in place of fins. Twenty inches long. A-, B-, C-, D-, or E- powered.

Scale Kits

Black Brant II. This scale model of the Canadian sounding rocket is an impressive 41.5 inches long and designed for use with high-powered FSI F engines. Nose cone, fins, and boat-tail are precut.

Black Brant III. This Estes scale model (Fig. 1-33) is based on another version of the Canadian rocket. Kit features include plastic nose cone and nozzle and decals. Easy to assemble, this rocket is capable of altitudes over 1000 feet with C-power.

Boeing A.C.L.M. Cruise Missile. Offered by Estes, the

Fig. 1-26. The Maxi-Alpha III, an enlarged version of Estes' Alpha III.

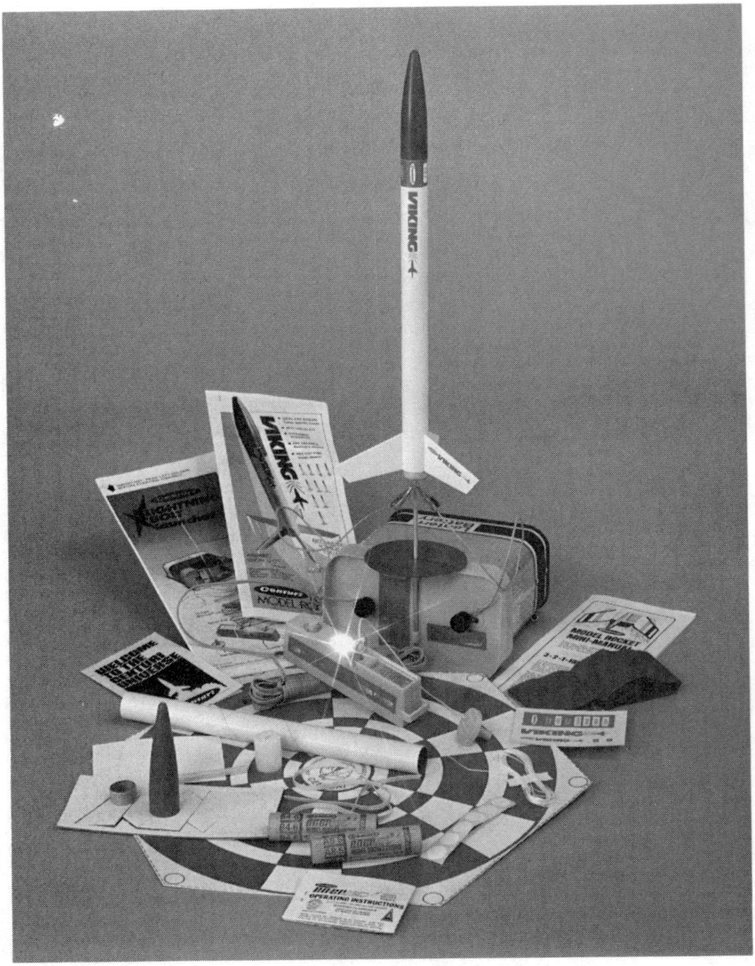

Fig. 1-27. The Viking from Centuri, an excellent beginners' kit.

Cruise Missile features authentic decals and deployable wings for display (Fig. 1-34). The body is blow-molded plastic, the fins and wings are die-cut plastic, and additional parts are injection-molded plastic. The model is quite a challenge to build, A-, B-, or C-powered.

LTV Scout. This is Estes' scale model of a four-stage solid propellant orbital launch vehicle. Thirty-six inches in length, this kit has many detailed plastic parts (Fig. 1-35). C-powered.

Magnum Jayhawk. Impressive 1/5 scale model of the United States Navy's AQM-37A target drone, from Centuri. Definitely an advanced kit, the Jawhawk is 30.4 inches in length and D-powered.

Mercury Redstone. A fine scale model of the historic launch vehicle that put the first American into space. Features a molded plastic Mercury space capsule and escape tower. Twenty-nine inches long, C-powered.

Phoenix. Estes' semiscale model of the air-to-air supersonic missile. Thirty inches long, 2.6 inches in diameter, the Phoenix (Fig. 1-36) is D-powered.

Saturn V. A favorite for experienced scale modelers, the Saturn V from Estes is 43.5 inches long and features corrugated wrap-ons, dummy engines for display, and many plastic parts (Fig. 1-37). Can be flown with a cluster of three C engines or with one D engine.

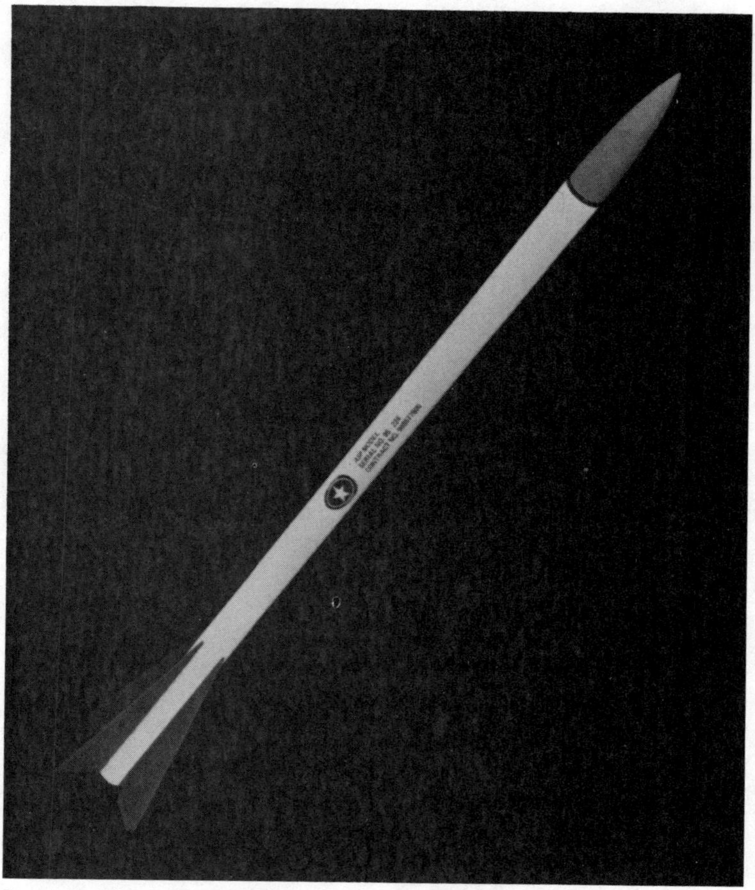

Fig. 1-28. The A.S.P. from Estes, a simple scale model.

Fig. 1-29. Space Shuttle Columbia, a semiscale kit from Estes.

Space Shuttle. This Estes Space Shuttle (Fig. 1-38) is scale in appearance and flight. A vaccu-formed plastic shuttle is attached to the external tank and solid rocket boosters, and the entire assembly lifts off under the power of one C engine. At apogee, the shuttle separates and glides back to Earth, while the rest of the rocket is recovered via parachute.

SR-71 Blackbird. Centuri's model of the U.S.A.F.'s secret high-altitude reconnaissance aircraft. Twelve-and-one-quarter inches in length, D-powered.

LEARNING MORE

As you become more involved in model rocketry, you will probably want more model rocket kits, materials for constructing your own model rockets, additional and perhaps bigger model rocket engines, and a more complete GSE.

In addition to reading this book and obtaining the aforementioned basic necessities, many other activities will help you become better acquainted with the hobby of model rocketry.

Obtain the catalogs of as many of the model rocket manufacturers as possible. (You'll find manufacturers' addresses in Appendix A of this book.) Study these catalogs and compare the products of the various manufacturers. Note the types of kits in which each manufacturer specializes, as well as the types and sizes of materials used, such as body tubes, nose cones, etc. All of this information will be very useful as you undertake various model rocketry projects.

Visix hobby shops and department stores in your area to look over the model rocket kits and supplies they offer. This will enable

Fig. 1-30. Centuri's Satellite Killer and Red Eye, companion kits.

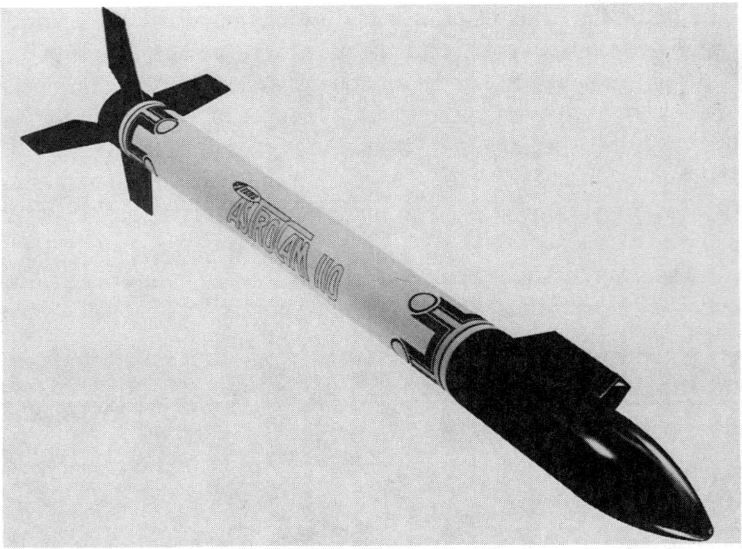

Fig. 1-31. Estes' Astrocam, the only model rocket-carried aerial camera.

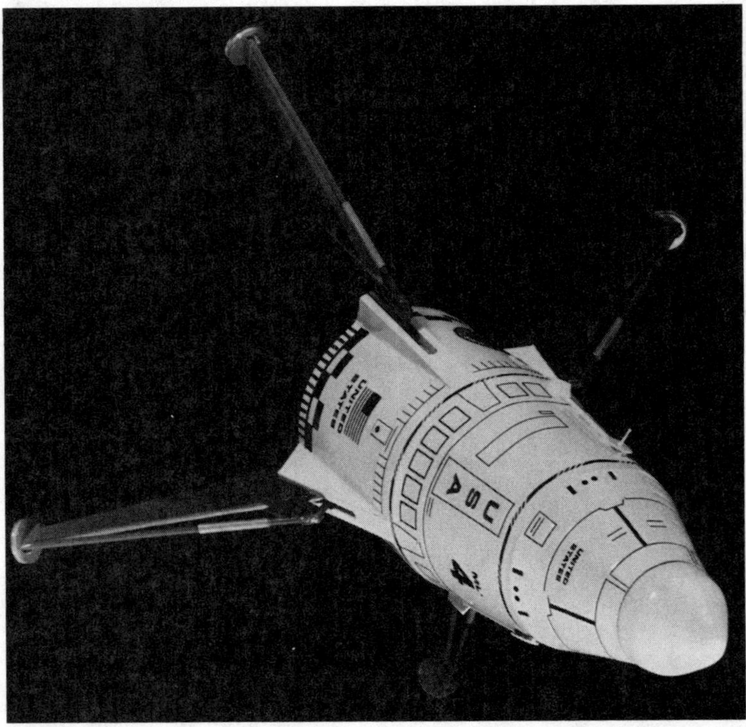

Fig. 1-32. The Mars Lander, a unique kit requiring a high degree of skill, from Estes.

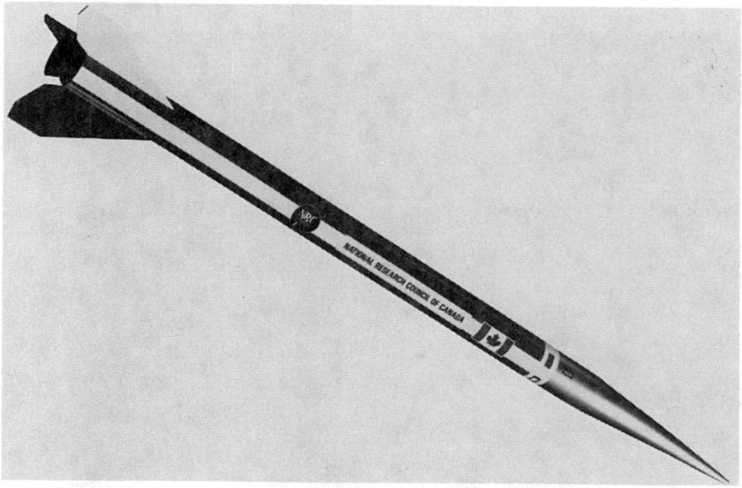

Fig. 1-33. Black Brant III, Estes Industries' scale model of the Canadian atmospheric sounding rocket.

Fig. 1-34. Another excellent scale model from Centuri—the Boeing A.C.L.M. "Cruise Missle."

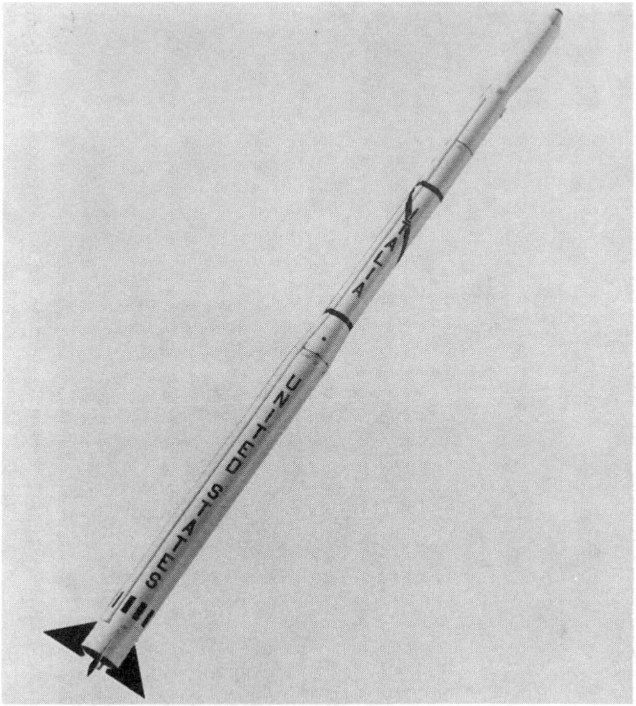

Fig. 1-35. The LTV Scout, a scale offering from Estes.

Fig. 1-36. The Phoenix—a D-powered model of the supersonic air-to-air missile, from Estes.

Fig. 1-37. The Saturn V Moon Rocket, one of the most popular scale kits, from Estes.

Fig. 1-38. Estes Industries' space shuttle—a model with both scale model appearance and performance.

you to make a more direct comparison of the offerings of various manufacturers and find out just what is available locally.

Check your area for large, open fields, including football and baseball fields, parks, and farm fields. This should be high on your list of priorities, because no matter what types of model rockets you plan to fly, you'll need a large, open area for flying. Once you've surveyed some likely fields, contact the owners of the fields, or those responsible for them, and ask permission to use the fields. (Never use a flying field without permission!) If you take the time to explain just what model rocketry is, you will usually find people receptive to letting you use a field.

Check for model rocketry clubs in your area. Local schools or civic groups may sponsor model rocket clubs. You may also find out about clubs through your local hobby shop, or by contacting the National Association of Rocketry.

Join the National Association of Rocketry. This will put you in direct contact with several thousand model rocketeers in all parts of the country, through a monthly magazine, local and regional clubs and activities, and various publications.

Spend some time at your local library and bookstore, searching for other books on model rocketry, as well as books on aeronautics and full-scale rocketry. Any additional background on rocketry and modeling will be helpful.

Chapter 2

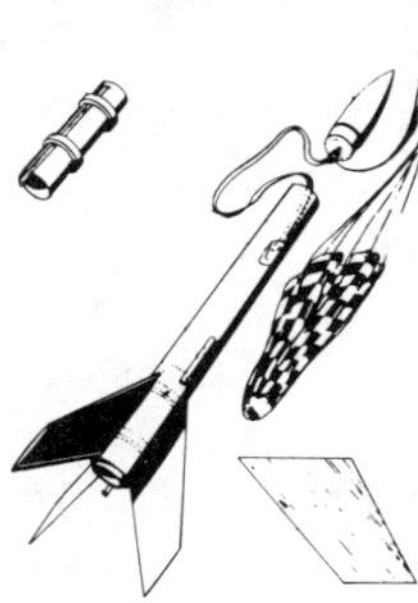

Brief History and Chronology of Rocketry

The rocket has been around a lot longer than most of us realize. The earliest recorded use of rockets is in the 10th century: the Chinese Emperor employed groups of artisians to produce powder burning rockets for use in the battlefield, and there is evidence that rockets were in use earlier.

HISTORY

In the mid-13th century, trade expeditions spread the word of rocketry to Europe and during this same period Arab armies were employing rockets against their enemies on a limited basis.

Although rockets were sometimes used by the Chinese for fireworks displays, the predominant application of rockets was in warfare. (It is interesting to note, in this context, that the name *rocket* is derived from the Italian word *rochetta*, which means "lance cover.")

There was almost no development of the rocket beyond its primitive form as an explosive-bearing projectile until the 1500s. At this time, theorists began designing rockets for applications other than warfare. *Leonardo da Vinci* sketched and described a two-stage rocket for attaining high altitudes. A Chinese inventor designed a rocket-powered kite to carry a man aloft. German and Italian designers planned a primitive version of the torpedo—rocket-powered, of course—and even a rocket message carrying device.

During the next 200 years, rocketry was slowly refined. Sir

Isaac Newton formulated his Laws of Motion, and improvements such as fins and tandem engine placements were proposed. Rockets were very popular with European royalty (as fireworks displays), and there were some experiments that involved sending animals aloft by rocket and recovering them via parachute.

Rockets were not considered a major element of warfare until the early 1800s, when *Colonel William Congreve*, a British military officer, began developing a rocket arsenal for Britain (this was initiated by the fact that the British had suffered devastating rocket attacks in India during insurrections).

Congreve proved that rockets fired from ships could usefully supplement cannon, and the British Navy used rockets with great success during engagements in India, Denmark, and elsewhere. Rockets figured prominently in the War of 1812, also, when the British bombarded Baltimore's Fort McHenry (Francis Scott Key was present during this bombardment and immortalized "the rockets' red glare" in our national anthem). Rockets also saw action in the American Civil war, and became a part of the artillery of virtually every army.

The principles of rocket space flight were developed by Russian scientist *Konstantin Tsiolkovsky,* at the turn of the century. Rockets has been envisioned as a method of air and space travel before this time by fiction writers such as Jules Verne, but Tsiolkovsky's work was the first practical and serious approach to the topic.

Tsiolkovsky continued his research and speculation on into the early 20th century. At this same time, *Robert Goddard*, working at Clark College in Massachusetts, began practical work on multistage solid and liquid fuel rockets. He was awarded several patents. In 1926, he launched the first liquid fuel rocket.

Interest in rocketry began mushrooming after Goddard's success with early rockets. Rocket societies were formed in the United States, Europe, and Russia, and the state of the art of rocketry was advanced quite a bit through the efforts of such societies. Science fiction was booming in the pulp magazines of the era, and the public became quickly acquainted with rocketry over the next decade through the adventures of comic book heros such as Buck Rogers and Flash Gordon.

During the two decades preceding World War II, experimenters in Germany undertook extensive activities in rocketry. *Hermann Oberth*, a theorist and peer of Goddard, inspired *Fritz Opel*, the automaker, to build and fly a rocket-powered glider, thus marking the first rocket-powered flight by a human. Opel also experimented with rocket-propelled cars. The first film to realistically depict space travel was produced in Germany during this period. *Woman in the Moon* was produced by the famous filmmaker, *Fritz Lang*, in 1929, the Oberth as technical advisor.

All of this set the stage for further German developments in rocketry in the late 1930s. *Werner von Braun* and other rocket enthusiasts, backed by the German army, eventually developed the infamous V-2 rocket, upon which the post-war rocket technology of both the United States and Soviet Union was based.

At this point, rocketry had come a full circle. Starting as crude paper and bamboo tubes filled with gunpowder, rockets had slowly progressed into areas outside of warfare. The potential for the use of rockets in space travel was beginning to hold promise. With the advent of World War II, however, all rocketry development focused on warfare applications.

After World War II, research into peaceful applications of rockets resumed. Von Braun and his counterparts in the Soviet Union raced to achieve superiority over the United States in rocketry, with each country having an eye toward placing a satellite in orbit for the International Geophysical Year in 1958. In the meantime, public awareness of the potential for rocketry in spaceflight continued to blossom, and research into military applications of rocketry continued.

The Soviets placed the first satellite in orbit, in 1957, and the space race was on. The next decade saw more advancement in rocketry than had the preceding 500 years.

The race inspired the first model rockets, and the growth and development of model rocketry parallel that of real rocketry, reaching a zenith in the early 1970s. Interest in spaceflight was at an all-time high during this period, thanks to the United States' highly successful space efforts.

Following Apollo and the manned lunar missions, there was a 10- year lull in space activity and public interest waned. Model rocketry stayed alive, of course, and the hobby experienced some growth. Today, with the space shuttle missions reawakening interest and involvement in space, model rocketry continues to grow and develop.

CHRONOLOGY

200 B.C.-200 A.D. Gunpowder formulated and firecrackers

used during the Chinese Han Dynasty.

1000-1200 A.D. The potential in warfare for the explosive properties of gunpowder is realized in China, and grenades, flame throwers, cannon, and rockets are developed.

1200-1250 A.D. Gunpowder and rocketry spread through the Arab world and into Europe, thanks to trade routes.

1500-1600 da Vinci describes the two-stage rocket; the first multistage rockets are flown in Romania by Conrad Haas. Other advancements—in theory and practice—are made in China and Europe.

1700 Rockets are common throughout Europe as a part of fireworks displays.

1804 Colonel William Congreve demonstrates the practicality of ship-launched rockets and helps establish Britain's arsenal of rockets.

1865 Jules Verne's *A Trip to the Moon* is published, the first science fiction about space travel. Although the spaceship in the story was launched from a cannon, it used rockets for maneuvering.

1903 Konstantin Tsiolkovsky proposed liquid-fuel rockets for space travel.

1922-1923 Early rocket societies formed; Hermann Oberth pioneers rocket theory.

1926 Robert Goddard launches first liquid-fuel rocket in Massachusetts.

1928 Fritz Opel tests rocket-powered autos and achieves the first manned rocket flight in a rocket-powered glider.

1942 Werner von Braun, head of the German military rocket development program, test flies early V-2.

1945-1956 Development of rocketry for both war and peace continues, largely carried out by former German rocket scientists in the United States and Russia.

1957 The world's first satellite—Sputnik—is placed in orbit by the Russians. The Space Age (and the space race) begins. The hobby of model rocketry is born.

1960 First weather satellite, Tiros 1.

1961 First manned spaceflights by Russia and the United States.

1969 First manned landing on moon.

1973 Skylab placed in orbit—first space station.

1977 Viking on Mars. 100 millionth model rocket flight.

1981 Space shuttle Columbia's maiden voyage is an unqualified success. More flights follow.

1982 Columbia continues test-flights. Russian cosmonauts make frequent use of *Salyut* space station, establishing space endurance records. The European Space Agency's *Ariane* is successfully flown, marking the first major space effort not involving the United States or Russia.

1983 Second space shuttle, the Challenger, is tested.

Chapter 3

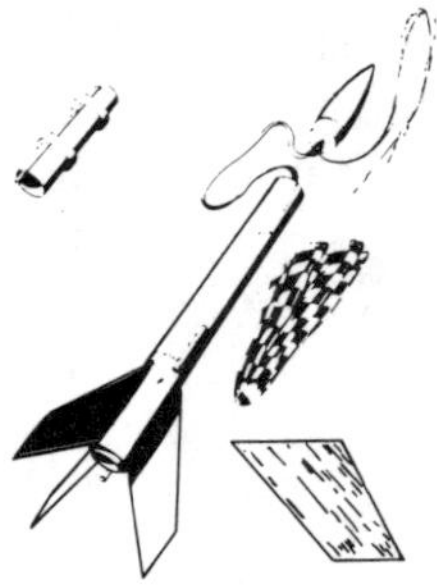

Construction and of Model Rockets Preparation

The basic modeling tools and materials (Fig. 3-1) needed to build model rockets include the following:

- X-acto modeling knife (or a single-edge razor blade).
- Fine sandpaper.
- Metal-edged ruler.
- Scissors.
- Pencil or marker.
- White glue (Titebond, Elmer's, or other brand).
- Paint brushes.
- Sanding sealer or dope (Aerogloss or other brand).
- Paint (spray enamel, butyrate dope for brush work).
- Masking tape.
- Cutting surface (heavy cardboard or "kevlar" sheet).

With all of these materials assembled, and a roomy, well-lighted work area, you are ready to build model rockets.

If you've never built a model rocket before, it is best that you built six or more kits of the various types available from model rocket manufacturers. Building kits will teach you a lot about how a model rocket should be designed and built and the best construction techniques to employ (Fig. 3-2).

It is a good choice to select one of the "Skill Level 1" kits, such as the "Alpha" (available from Estes Industries), or Centuri En-

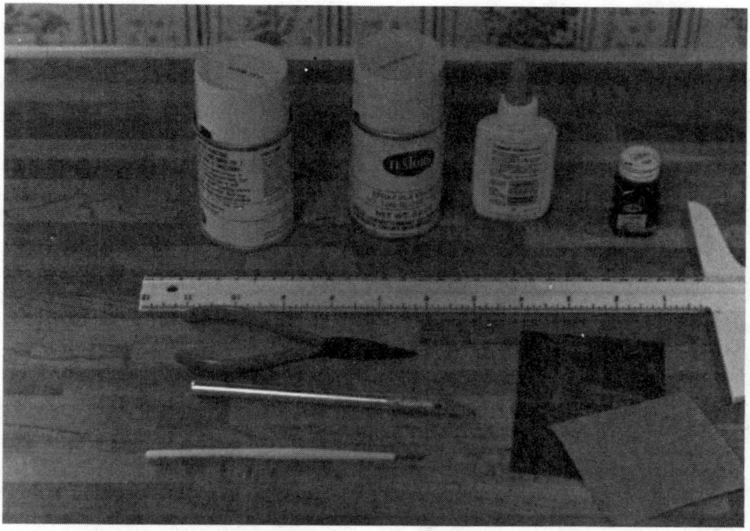

Fig. 3-1. Typical model rocket building supplies include glue, paints, razor knife, sandpaper, measuring stick.

gineering Company's "Screaming Eagle." Simple beginner's kits are also available from Flight Systems, Inc. and Canaroc Spacemodels, but the Estes and Centuri models are more easily found in your local hobby shop or department store.

After building two or three "Skill Level 1" kits, you should feel confident enough to move on to higher skill levels. Manufacturers' catalogs usually rate the degree of difficulty of their kits, so you can use these as a guide.

You may also find "multiple purpose" kits to be of help after you've progressed beyond the beginner's stage. Centuri's "Power System Outfit" is an excellent learning system, providing the

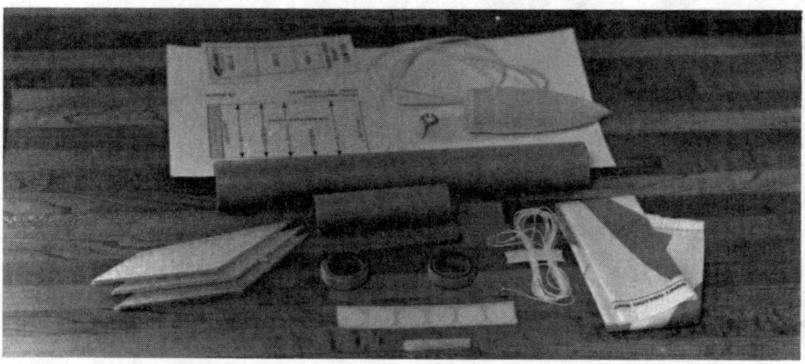

Fig. 3-2. An unassembled kit contains all you need to construct model.

modeler with low- and high-power rockets, as well as multistage models. Estes Industries offers a "Multi-Roc" kit, which provides single- and two-stage capability, as well as a payload section and a boost-glider—all in one kit. Kits of this type provide an easy introduction into advanced model rocket activities. (See Chapter 1.)

TIPS FOR FIRST-TIME BUILDERS

The first thing to do when building any kit is to lay all of the parts out carefully and identify them with the parts list and blown-up diagram accompanying the instructions for assembling the kit.

After identifying the parts, you should read the instructions through completely, at least once (Fig. 3-3). This seems elementary, but it is necessary, because jumping ahead of the instructions and assembling parts that "obviously" go together can leave you with a model that is improperly assembled and will never fly. Junk, in a word.

After studying the directions, you should begin with Step 1 of the instructions and follow them closely. Assembly instructions are almost always clear and easy to understand, but if you encounter any problems with assembly instructions, you should make a *dry-run assembly*. This is done by assembling the rocket, per the instructions, but without using any glue. Doing this should clear up any problems for you.

As you are building the model, refer frequently to the blown-

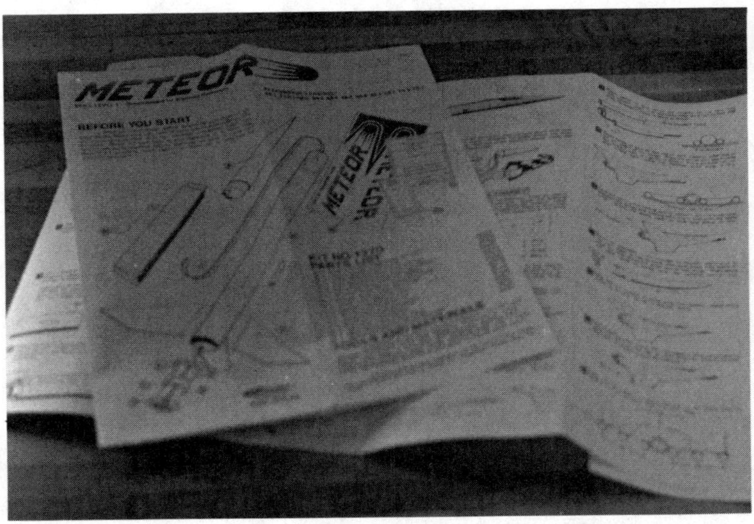

Fig. 3-3. Detailed instructions are always included with a kit. Follow them closely.

up diagram of the rocket's parts, which will show you how the rocket goes together as a whole. Also check the appearance of your rocket against the photo or drawing of the finished product that accompanies the instructions.

In every step, be certain that you have it right, because once the glue has set on the parts, you will find it extremely difficult (and usually impossible) to separate the parts and correct your mistakes.

ASSEMBLY

Balsa, in addition to being lightweight and strong in relation to its weight, is extremely easy to work with. These factors make the wood ideal for use with model rockets.

Balsa is easily cut into any shape with an X-acto knife or razor blade. Because a blade goes through the wood so easily, you should take care not to apply too much pressure to a knife or blade when cutting balsa. This will prevent the blade's slipping and causing a ruined sheet of balsa or a bad cut on your finger. You should also have a cutting sheet under the balsa with which you are working.

When cutting balsa fins from sheet balsa, you should make certain that the grain of the balsa does not run parallel with the root edge of the fin, as illustrated in Fig. 3-4. If the grain is parallel to

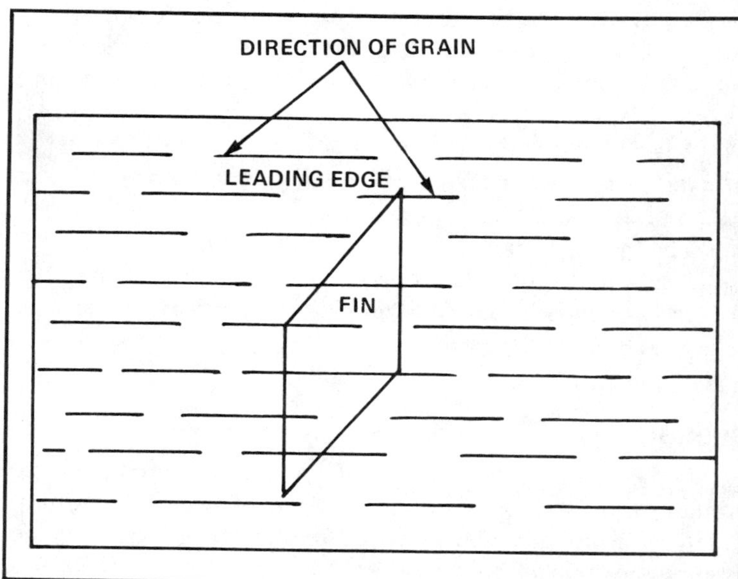

Fig. 3-4. Cutting a fin from balsa stock.

Fig. 3-5. An assembled model rocket made from a kit.

the root edge of the fin, the fin will most likely break the first time the rocket is flown, or on recovery.

When gluing parts together, it is best to pre-position parts to ensure that they will fit together properly (Fig. 3-5). Once you are satisfied that the fit is right, apply a thin layer of glue to one of the two parts to be joined (such as a fin) and touch it to the other part briefly. Separate the parts, and apply more glue to both, giving the glue a minute to set up. After the glue is tacky to the touch, put the two parts together and hold or clamp them until the glue is dry.

(A useful alternative to holding or clamping fins is the use of a *fin jig,* such as the one pictured here (Fig. 3-6). A fin jig will also assure perfect alignment of fins.)

Some parts, such as fins and launch lugs, may be reinforced with beads of glue to make "*fillets,*" as in Fig. 3-7. You may also use strips of paper or fabric to provide additional reinforcement, as in Fig. 3-8.

GLUES

In the past few years, dozens of new adhesive products have hit the market, and there will likely be many more in the future. This section will give you a brief description of the types of glue used by most model builders.

White Glue. A water-based, synthetic adhesive. Works great

Fig. 3-6. A fin jig. Courtesy Estes Industries.

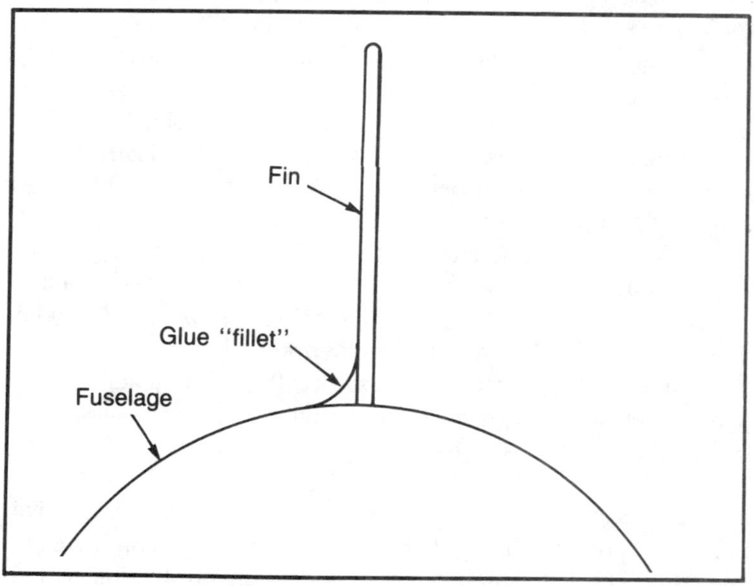

Fig. 3-7. A glue fillet will add strength to a fin.

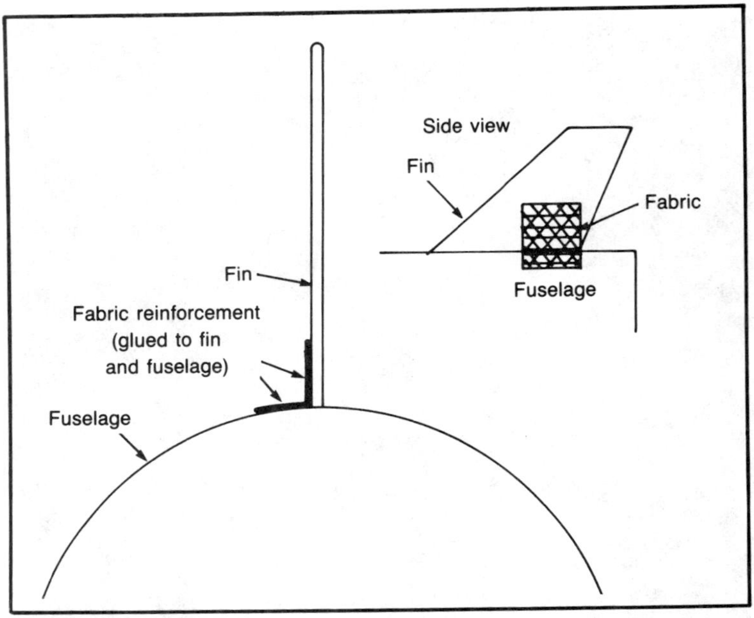

Fig. 3-8. A strip of fabric can be glued to a fin for added strength.

on wood, paper, and cardboard. Dries clear and reaches full bonding strength in 48 to 72 hours. White glue is nontoxic, nonflammable, has little odor, and the fumes are harmless.

Disadvantages include a relatively long set-up time and water-solubility.

Brand names include "Elmer's" and "Will-Hold."

Yellow Glue. Very similar to white glue, but specially formulated for gluing wood. It penetrates wood better than white glue and forms stronger joints.

Yellow glue also sets up quicker than white glue and is more easily sanded. Yellow glue is more heat- and solvent-resistant.

Brand names include "Elmer's Carpenter's Glue," "Franklin Titebond," and "Weldwood Wood Worker Glue."

Epoxies. Very strong and versatile adhesives that adhere well to most metals, wood, glass, ceramics, and some plastics. They have excellent resistance to solvents and can take temperatures up to 150-200 degrees Fahrenheit.

Epoxies divide into two categories: *regular* and *quick set* (or "five minute") epoxies. The regular epoxy sets in about 30 minutes to one hour. The "five minute" epoxy sets up in 4 or 5 minutes and reaches full bonding strength in under 24 hours.

One disadvantage with epoxies is the fact that they don't "grab" body tube exteriors and balsa fins as well as white and yellow glues. Epoxy is better used on interior components such as engine mounts and shock cord mounts, due to its excellent strength.

("Five minute" epoxy is also great for setting fins. Put fin holes in the fin line on the body tube and then place a small amount of epoxy on the line and on the root edge of the fin. Let the epoxy cure for 3 minutes and then attach the fin. The fin will set in 5 minutes, and the epoxy will have the same effect as miniature bolts in holding the fin in place. Let the rocket set overnight after you've attached all of the fins in this way.)

Popular epoxy brands include "Duro," "Fox," and "Devcon."

Instant Glues. Sometimes known as "super glue" instant glues are *cyanoacrylates*. They set up on most nonporous surfaces in less than one minute. These glues are excellent for gluing plastics and attaching fins quickly.

The disadvantages of glues of this type include dangerous fumes and the fact that even a small amount can glue skin almost instantly. Handle "super glue" with extreme caution!

Duro, Devcon, and other companies make this type of glue.

Plastic Glues. In addition to the instant glue described above, there are three types of adhesives designed for plastics:

Solvent glue, a clear liquid that actually fuses plastic pieces together. Highly recommended for precise plastic model work. "Plastruct" and "Testors" are popular brands.

Model cement, also known as "airplane glue," this type of glue is used to bond polystyrene and acrylic plastics. This comes in tubes, and you've probably used it to build model cars, planes, or ships.

Model cement sets in about 10 minutes. Its major disadvantages include the fact that it is usually thick and difficult to work with, it will quickly "eat" any plastic it comes into contact with, and it gives off dangerous fumes that should not be inhaled. It is also flammable. Handle with care, and use only in a well-ventilated area. Available in any hobby shop from Testors and a variety of other manufacturers.

Vinyl menders, designed for bonding vinyls and rigid plastics, such as acrylics, phenolics, and polystyrene. Used mainly in the manufacture and repair of vinyl and plastic products and not readily available.

SANDING AND SEALING

All balsa parts of a model rocket should be sealed with several layers of sanding sealer. The balsa must first be sanded very smooth with fine sandpaper, so that any cracks or pits in its surface can be eliminated. Sandpaper, or a nail file, should also be used before sealing to sand down the leading and trailing edges, and the tips of fins to a streamlined shape, as in Fig. 3-9.

After all sanding and shaping is completed, a medium layer of sanding sealer should be applied to the surfaces of the balsa parts, and allowed to dry. When the sealer is completely dry (hard to the touch), it should be sanded smooth, and a second layer should be applied, allowed to dry, and sanded. This process should be repeated until you have a surface almost as smooth as plastic (three or four coats are required in most cases).

Carefully shaping, sanding, and sealing the balsa parts as described will greatly reduce aerodynamic drag and enhance your model's flight performance (Fig. 3-10).

You may also wish to add sanding sealer or a layer of white glue to the body tubes of your model. This should be done carefully, with a layer only thick enough to fill in the spirals on the body tube, and you will have to take extra care in sanding the body tube to ensure a uniform shape.

In all cases, the sanding, shaping, and sealing operations should be completed on each part, before it is attached to any other part

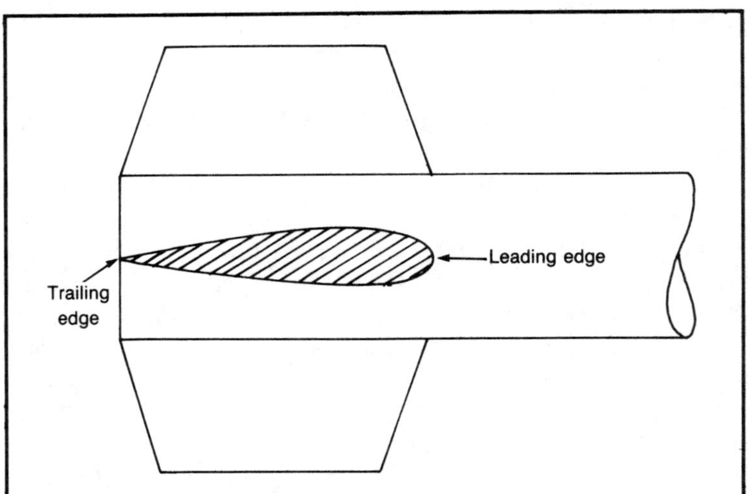

Fig. 3-9. Fin cross section. Courtesy Centuri Engineering Co.

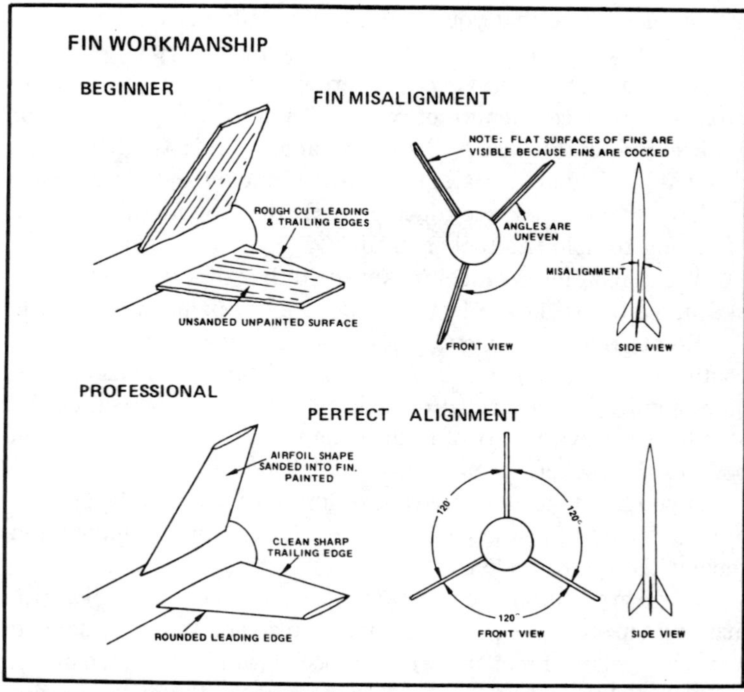

Fig. 3-10. Carefully crafted fins will be both more attractive and more durable. Courtesy Estes Industries.

(multisegmented fins should be assembled first, however). You should also avoid putting sealer on the surfaces of parts to be glued, such as the root edges of fins.

PAINT

After completely assembling your model, you will want to paint it. Enamel paints are best if you are spray painting, while butyrate dope is best for detailing with a paint brush. (You may also find *monokote*, which is used by model airplane builders as a wing and fuselage covering, useful in detailing. Simply cut small strips of the appropriate color, and apply them after painting.)

An airbrush may also be used, if you are very particular about the smoothness of the finish on your rocket. If you do not have an airbrush, but want an extremely smooth finish, I recommend using Krylon brand spray paints.

It is best to paint outside when using spray paints, unless the weather is excessively cold, wet, or windy. If you have to paint inside, make sure that you have a well-ventilated work area and that you cover with newspaper anything on which you don't want to get paint. When spraying an entire model rocket, the best approach is to place a length of rolled-up newspaper (Fig. 3-11), or a dowel rod, in the base of the rocket and use this as a handle. In this way, you can cover all of the rocket quickly and easily, without having to go back and touch up parts that you missed because of having to hold the rocket itself.

When painting a model rocket, hold the paint can motionless and move the rocket in front of the spray of paint, nose to tail.

Rotate the rocket slowly as you spray, making sure that the coat you are applying is even and smooth. The fins will usually require a quick touch up coat of their own, because you will not be able to reach every part of each fin as you are painting the main body of the rocket.

If you find that there are runs or lumps of paint on the surface of your rocket after the paint dries, sand these areas smooth and repaint the entire rocket.

If you are painting your rocket more than one major color, use masking tape to cover those areas that you do not wish to be painted one color (yellow, for example) and proceed with painting the model the primary color (Fig. 3-12). After the coats of the primary color have dried, use masking tape and newspaper to protect that color from the second color (red, for example) and spray the rocket with the second color.

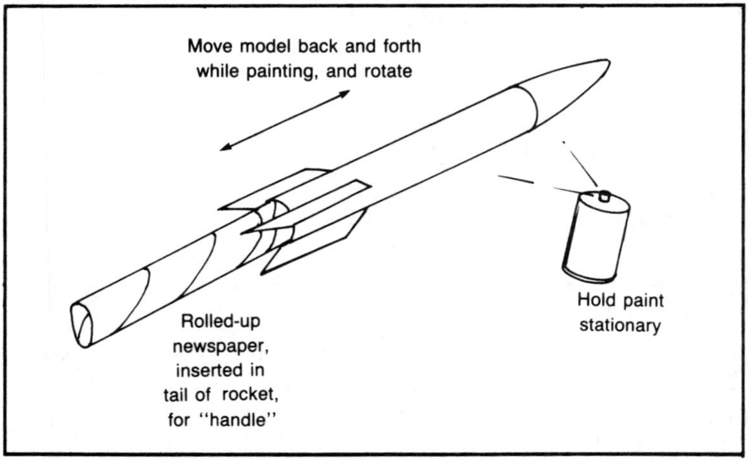

Fig. 3-11. Painting a model rocket adds realism.

You will find that it is best to use several coats of paint to turn out a nice looking model. It is also a good idea to start painting by applying a *base coat* of flat white paint, no matter what your final color or colors will be.

After you have completed painting your rocket, add any details, such as paint stripes or monokote shapes or strips, and allow these additions to dry. Next, apply any decals you wish to use, using a decal set product, which will make your decals adhere to the rocket more permanently.

All of this complete, you are ready to apply some protection to the finish and decoration of your model. The entire model should be sprayed with clear lacquer (either a clear gloss or a clear dull coat) to protect the finish from chipping and to make it easy to clean.

You may also wish to apply flameproof paint or coatings, such as those used by aircraft modelers, to the areas of the rocket that will be exposed to heat: the trailing edges of the fins, the base of the fuselage, and the top of the first stage and second stage if the rocket is a multistage model.

DESIGNING YOUR OWN ROCKETS

Much has been written about designing rockets. Centuri and Estes both offer complete manuals on this topic, but I will discuss the basics here. (Centuri and Estes also offer parts for designers, either individually or in "Designer's Special" packs (Fig. 3-13).

First and foremost, any rocket you design should be *stable*. This means that it should proceed in the direction it is pointed.

The two primary considerations when discussing stability are *center of gravity* and *center of pressure*.

The center of gravity (CG) is the point at which the rocket is in equilibrium, or the *balance point* of the rocket. The easiest method for locating the CG of a model rocket is to suspend the rocket in a loop of string, as in Fig. 3-14, and move it back and forth until it is balanced. The CG will be the point at which the rocket balances (equal weight ahead of and behind the loop). Should you need to change the location of a rocket's CG, all that is necessary is to add weight to the nose or tail of the rocket. Clay is a good material for adding weight.

The center of pressure (CP) of a model rocket is the point at which all aerodynamic forces center. There is as much aerodynamic force ahead of the CP as there is behind it. Like the CG, the CP

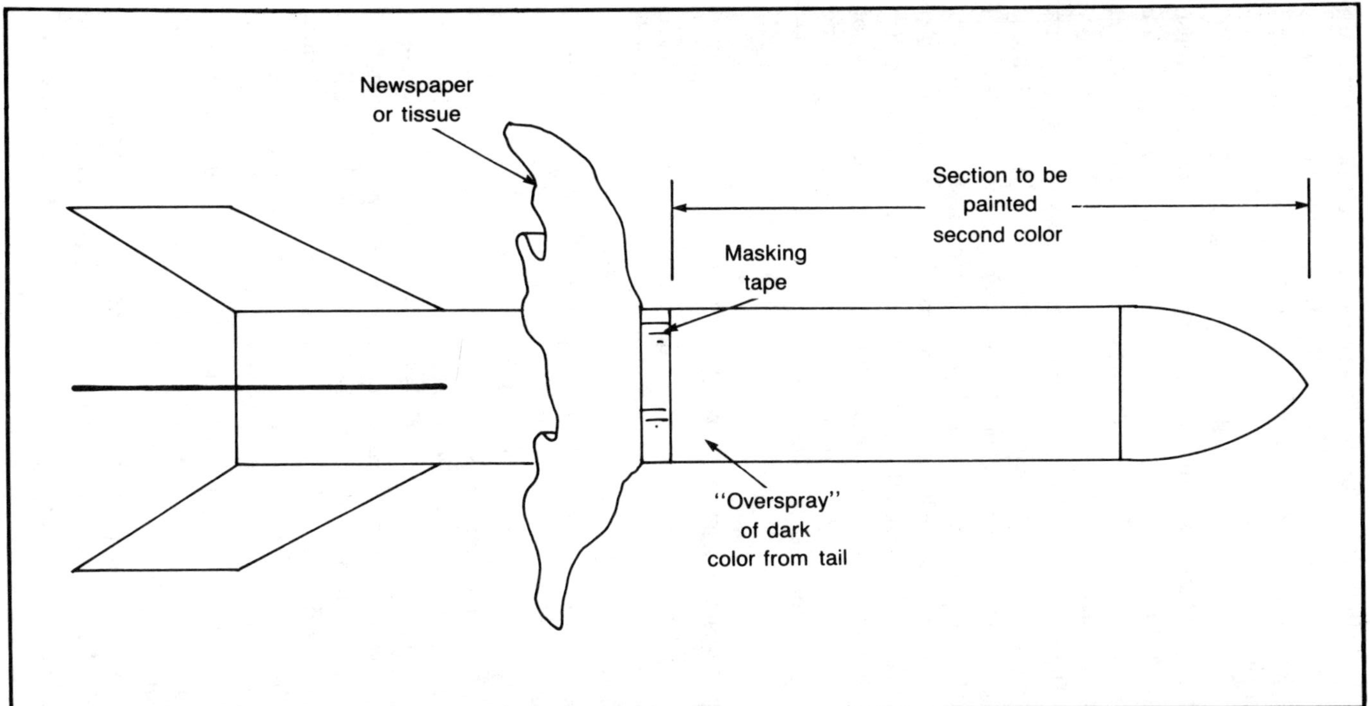

Fig. 3-12. Masking a model for two-tone painting.

Fig. 3-13. Prepackaged parts assortment for making model rockets of your own design. Courtesy Estes Industries.

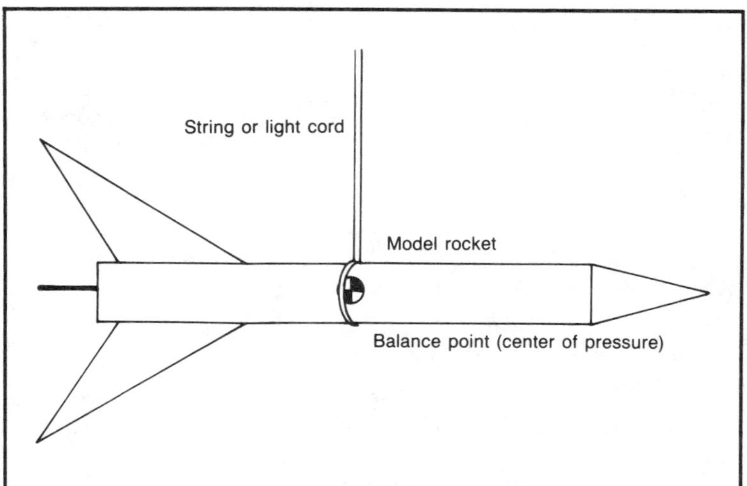

Fig. 3-14. String method for locating center of gravity.

is a balancing point for forces acting upon the rocket in flight.

To get a better idea of what CP is, look at the diagram of the model rocket in Fig. 3-15. Imagine that this drawing is a cut-out of a model rocket. Now, consider at what point you would have to hold the rocket to balance it against the force of a wind. This point will be the CP, and there will be equal surface area both ahead of and behind this point.

The CP is marked in this diagram with the symbol ⊙. The CG is marked with the symbol ◕. As you can see, the CP is more toward the rear of the rocket than the CG. This is because the fins of the rocket constitute surface area, surface area upon which aerodynamic forces can act. Obviously, the larger the fin area, the farther back the rocket's CP will be. Because the surface area ahead of the CP must equal the surface area behind the CP, adding larger fin area means that more of the rocket's fuselage will be ahead of the CP.

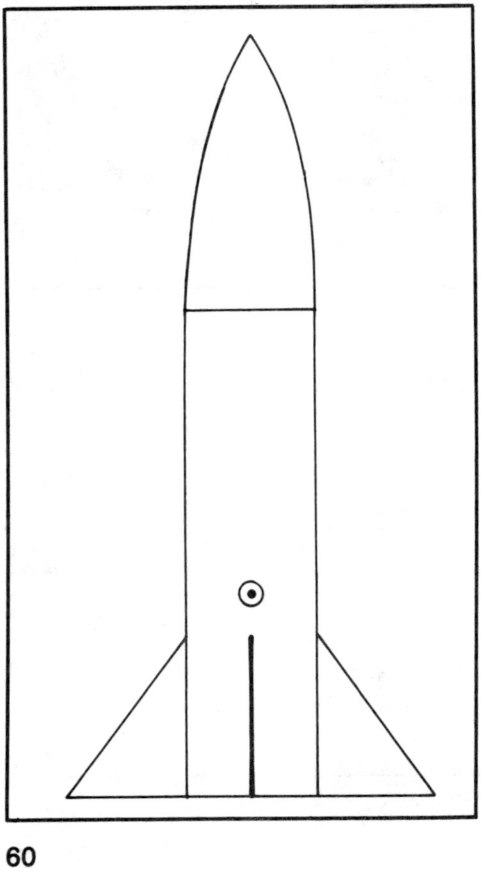

Fig. 3-15. The center of pressure on a rocket.

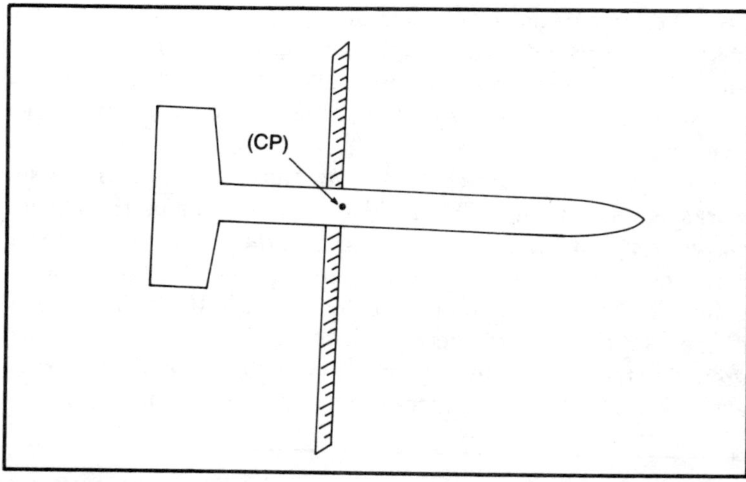

Fig. 3-16. Cardboard cutout method of locating center of pressure. Courtesy Centuri Engineering.

As you can see, changing the size of the fins is an easy way to change the location of a rocket's CP. Larger fins move the CP toward the rear of the rocket, while smaller fins will place the CP toward the front. An alternate method of changing the location of the CP is to lengthen or shorten the fuselage. The longer the fuselage, the farther ahead the CP will be.

The easiest method for locating a model's CP is the cardboard cutout method. With this method, you trace the outline of your model on a piece of thick cardboard and cut out the outline. Next, balance the cutout on a narrow edge, such as a ruler, as in Fig. 3-16. Because the cardboard is the same thickness throughout, the cutout will balance at a point at which there are equal areas on both sides. This point will be the CP.

An alternative method is to trace the outline of your model on a piece of graph paper and locate the point on the paper that has an equal number of squares enclosed by the outline on both sides.

GUIDANCE

Now that you have some understanding of what CG and CP are, let's take a look at what actually guides a model rocket and how CG and CP affect the guidance.

The fins of a model rocket keep the model flying in the path it is set in at launch. If the rocket begins to turn away from this path, the force of the air moving past the rocket will push on the fins until the rocket is again flying in the proper direction, as in Fig. 3-17. In other words, the aerodynamic forces acting upon the rocket tend to position it so that it presents the least area to the direction of flight.

In flight, a model rocket *oscillates* back and forth as the air works against the fins. This oscillation is centered on the CG, as shown in Fig. 3-18. The CP is, of course, the center point for the aerodynamic forces acting on the rocket, and the CG acts as a pivot point. The CP must therefore be behind the CG. If the CP is ahead of the CG, the nose of the rocket would be pushed toward the ground as the rocket is flying upward, since the aerodynamic forces act to keep the CP away from the direction of flight.

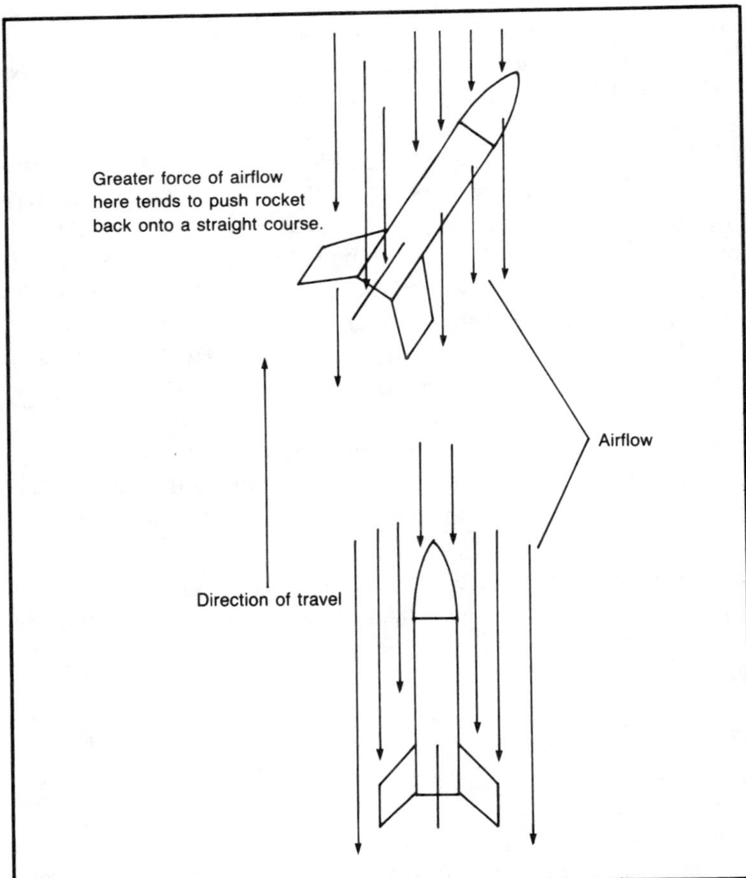

Fig. 3-17. Air flow acting on the fins of a model rocket.

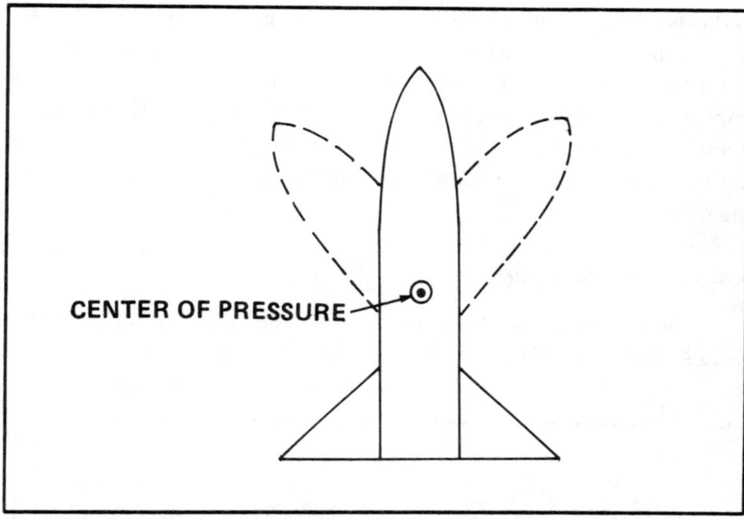

Fig. 3-18. Oscillation of a model rocket about its center of pressure.

Because of this, all model rockets rely on large rear fin areas for guidance and stability (Fig. 3-19). In this context, you should also be aware that forward fins, as pretty as they might look on a model, do not add to the stability of a rocket. Forward fins move the CP forward, because they add surface area ahead of the CG, and will cause the rocket to be unstable.

Fins that are too small in relation to the length of a rocket's fuselage can also result in instability. Very small fins will cause the CP to move close to the CG, perhaps even placing the two at the same location.

Extra large fins, in addition to adding useless drag and weight to a model rocket, move the CP back, away from the CG, and can cause violent oscillation during flight. This is due to the fact that the distance between the CP and the CG is like the distance between the end of a lever and the fulcrum; the force is increased by the increase in length.

The ideal model rocket has what is known as *one caliber stability*. This is achieved by making the distance between the CP and the CG equal to the diameter of the rocket's fuselage.

THE SWING TEST

The final test for stability before flight, and one which will give you a pretty definite determination of your model's stability, is the *swing test* (Fig. 3-20). The swing test is performed by attaching a string (usually with tape) to the model rocket's CG and swinging the rocket in a circle. In doing this, you will be able to approximate the forces acting upon the rocket in flight. If the rocket keeps its nose pointed in the direction you are swinging it, it is stable. If it oscillates violently, or turns completely around, it is unstable.

WEATHERCOCKING

The preceding discussion of the forces action upon a model rocket give the picture of an ideal flight. Nothing is ideal, of course, so we will consider one additional variable force—wind.

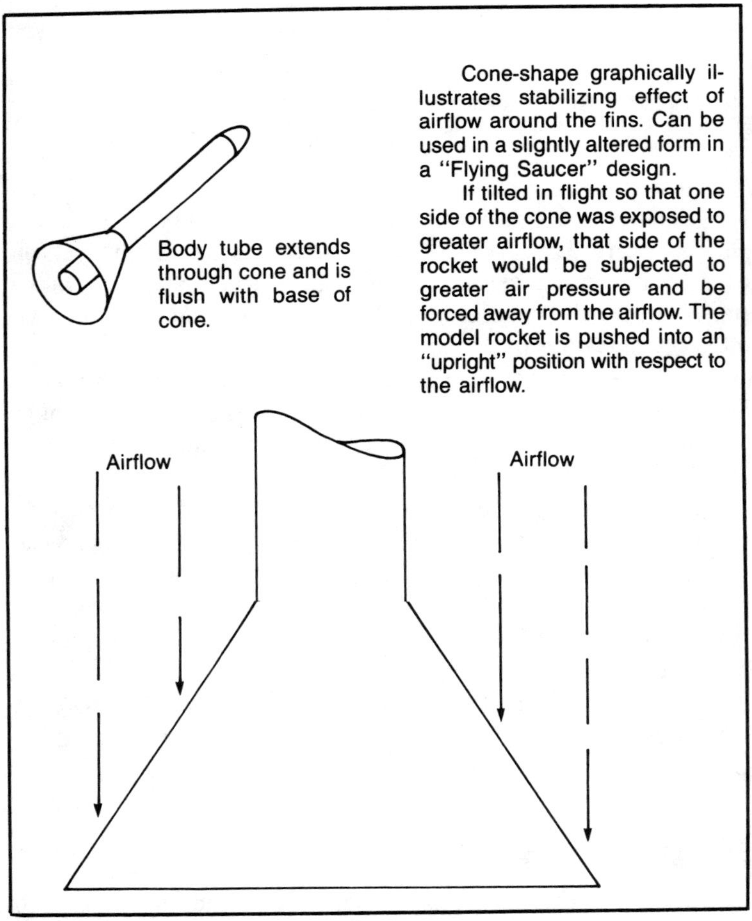

Fig. 3-19. An alternative method of stabilization.

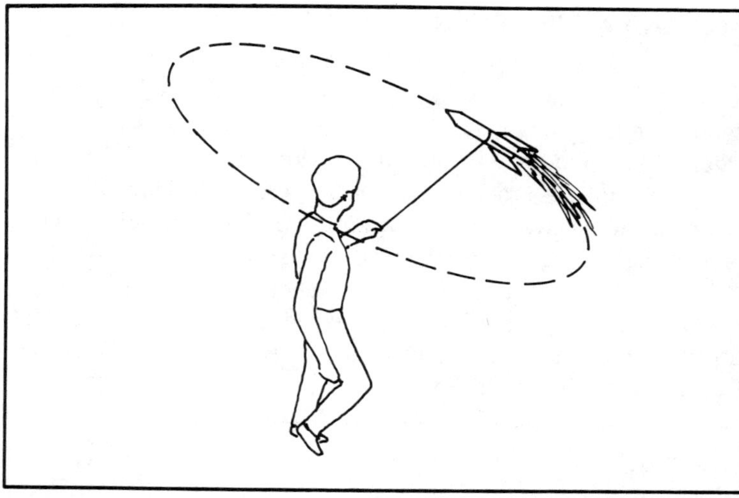

Fig. 3-20. Swing testing a model rocket.

As you have learned, the force of the air moving past a model rocket as it flies pushes against the fins so that the smallest possible area is presented to the direction of movement. If there is no wind, the rocket will continue in vertical flight (or in as near a vertical direction as it was launched). A small amount of wind, however, will push against the fins, turning them in the direction of flight. To compensate, the rocket will lean, or *weathercock* in the direction of the wind, until the two forces are equalized, just as a weather vane turns to the wind (Fig. 3-21).

The greater the wind, the more the rocket will weathercock. With enough wind, the rocket could turn until it is flying horizontally! Obviously, a smaller fin area will decrease the amount of weathercocking, but will also decrease the stability of your rocket, because the smaller fin area will permit the CP to move forward, toward the CG. This decrease in stability can be overcome by adding weight to the nose of the rocket (which will move the CG forward), but this in turn may cut down the performance.

Obviously, there are some trade-offs to be considered. If you expect wind during most of your flying, it is a good idea to alter the rocket with smaller fins so that it is stable. On the other hand, if you have a large flying field and don't mind running a good distance to retrieve rockets, you may want to chance weathercocking. You can also tilt the rocket on the launch pad to compensate for weathercocking and thus ensure a more vertical flight. The tilt should be away from the wind.

FINS AND NOSE CONES

Fins are necessary because they provide a model rocket with stability and guidance, but they also create *drag*. That is, their resistance to the movement of air causes the rocket to slow down. Figure 3-22 illustrates the various fin cross sections. The first—*streamlined*—cross section is the least drag-inducing. The rounded leading edge permits the air to flow easily past it, and the pointed trailing edge causes the least possible amount of turbulence. Cross sections two through four represent decreasing efficiency.

Fin shapes are very important in reducing drag. Figure 3-23 illustrates the most common fin shapes used in model rocketry, in decreasing order of efficiency.

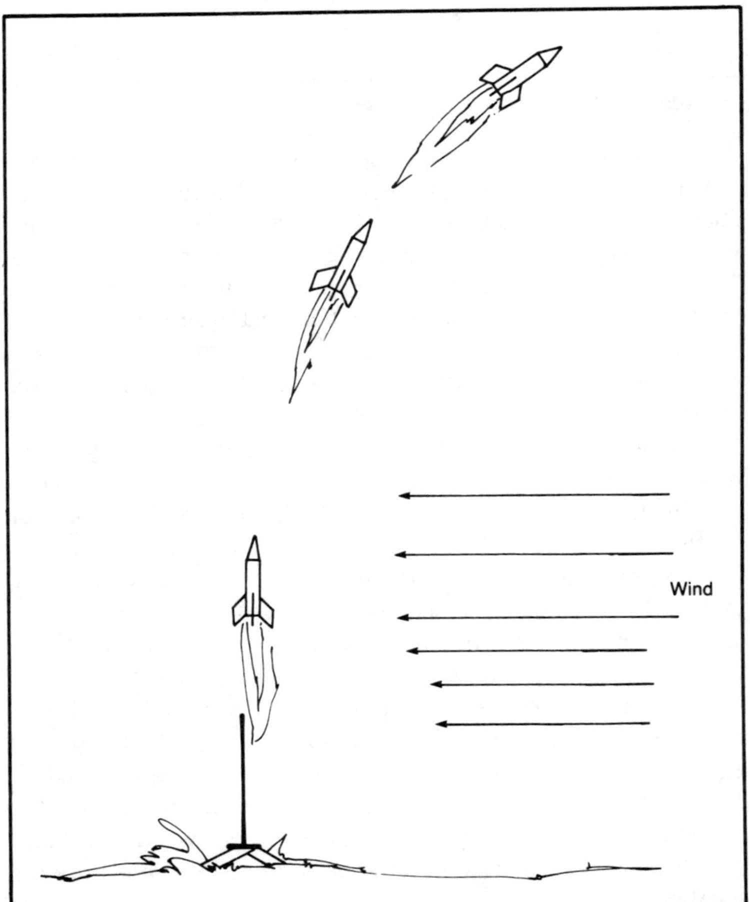

Fig. 3-21. Weathercocking into wind after launch.

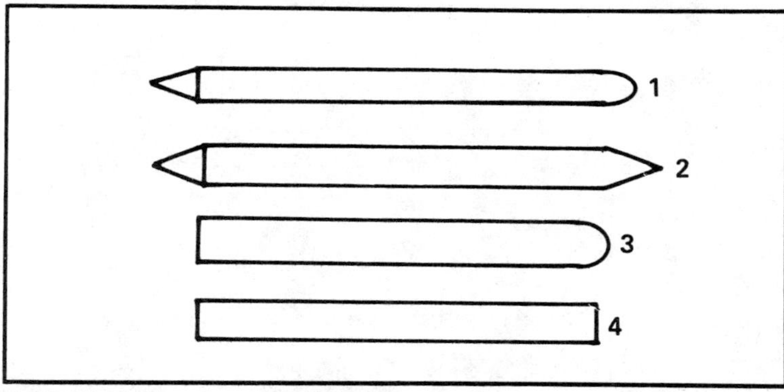

Fig. 3-22. Fin cross sections; poor (4) to efficient (1).

Nose cones can also cause drag (Fig. 3-24). Obviously, a flat nose will cause a great amount of drag during a model rocket's flight. A pointed nose cone will cause a lesser amount of drag, because it presents a smaller area (and less resistance) to the airflow. Figure 3-25 shows various types of nose cones, again in decreasing order of efficiency.

FLIGHT TESTING

After choosing the best design features (fin and nose cone

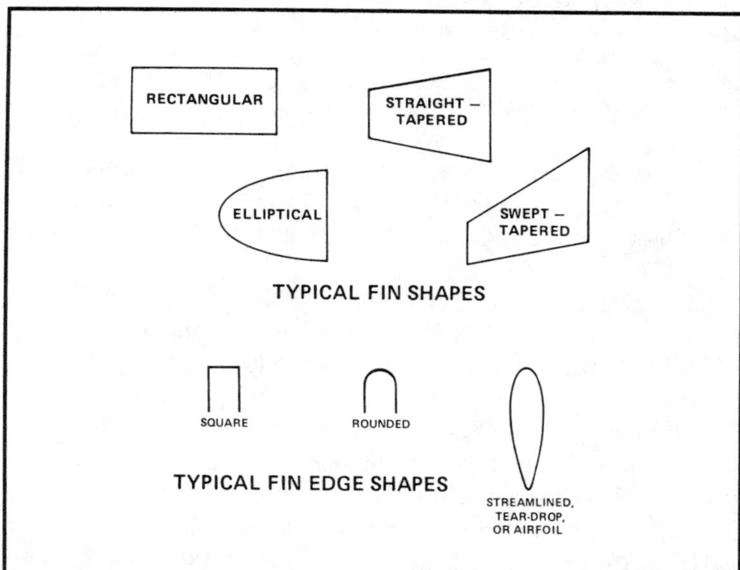

Fig. 3-23. Commonly used fin shapes. Courtesy Centuri Engineering Co.

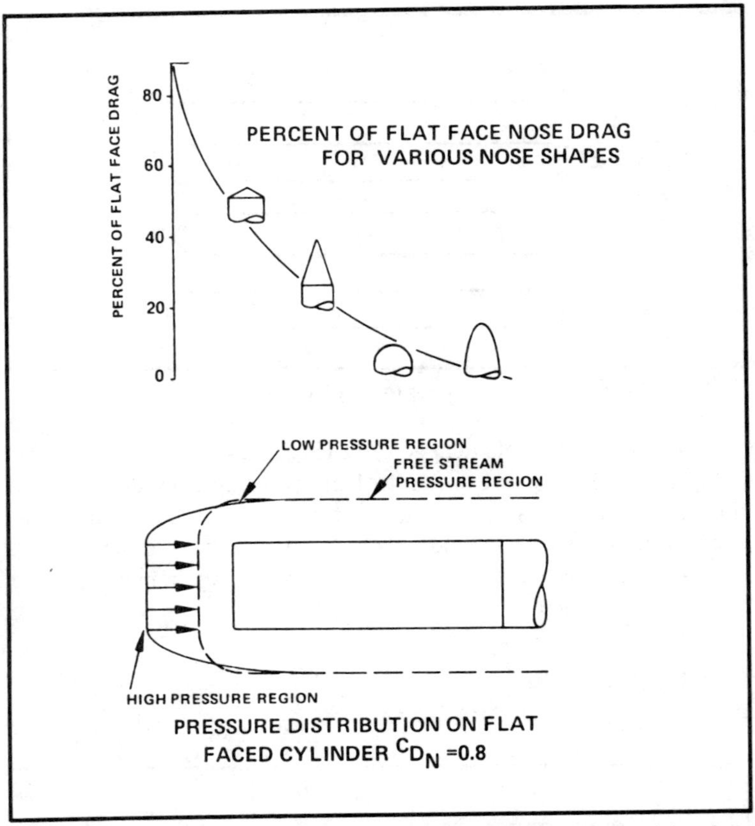

Fig. 3-24. The shape of a nose cone can alter the efficiency of a model rocket. Courtesy Estes Industries.

shapes, size, and power range), planning for stability, and constructing your own design carefully, you will want to flight test it. The first flight test of any untied design should be planned carefully, with an eye toward safety.

The size of your rocket will dictate the power needed to fly it. Obviously, a very small rocket will use mini-engines (A, 1/2A, or 1/4A). A medium-size rocket will fly with A, B, or C engines, and a very large rocket will need D or higher power. How do you decide which class of engine to use and what delay time to select?

The best way is to go through model rocket manufacturer catalogs and find several models of the same size, weight, and general design as yours, and see what engines are recommended for those models. After doing this, take into consideration your rocket's characteristics. Is it a bit heavier than similar models (perhaps be-

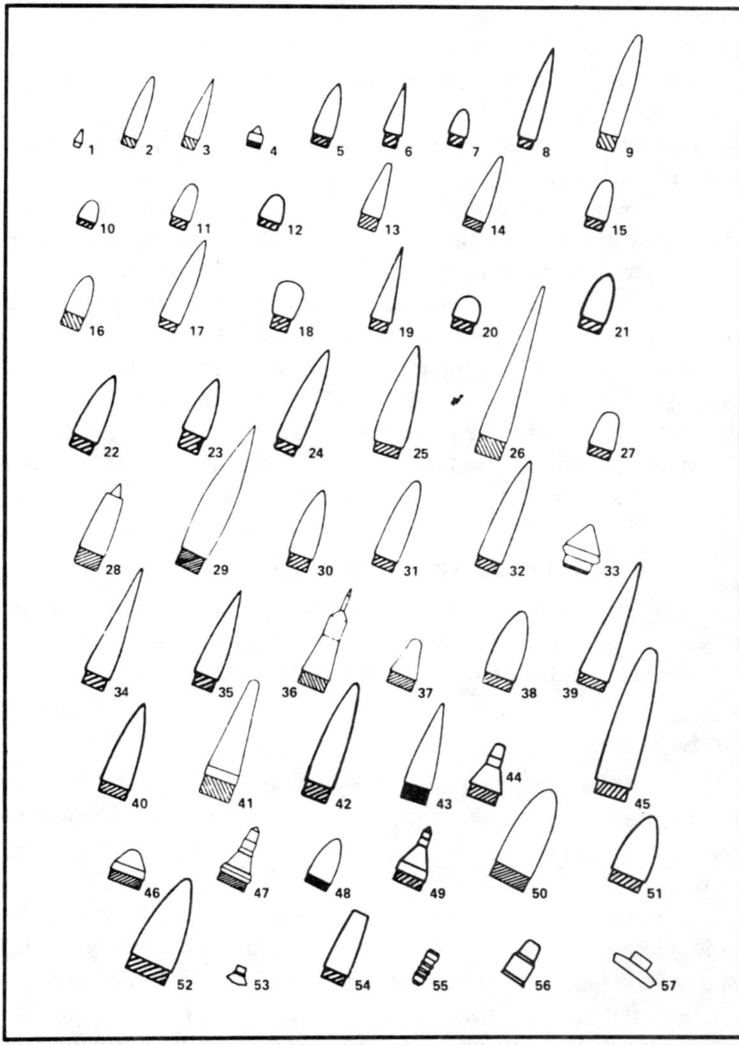

Fig. 3-25. Commonly used nose cone shapes. Courtesy Estes Industries.

cause you used small fins and added quite a bit of weight to the nose to move the CG forward?) If this is the case, you will want to use an engine with a higher average impulse (a B-6 or B-14 instead of a B-4, for example), and one with a short delay time, because the rocket will not gain much altitude before it begins to fall.

If, on the other hand, your model is lighter in weight and has less overall drag than similar models, you will want to use a longer delay time than that recommended by the manufacturer for similar models. This is because your model, if it performs to your expectations, will coast upward for a longer period of time and should be permitted to fall for a longer time to ensure recovery within a specified area.

Once you have selected the power for your rocket, you will want to consider which recovery method to use. A good rule of thumb is to use parachute recovery for larger, low-altitude rockets, and streamer recovery for lightweight, high-altitude rockets. Parachute recovery will provide a slow, safe descent, while streamer recovery will bring a rocket down quickly enough to keep it within the recovery area, while slowing it down enough to prevent damage.

The final consideration in testing new designs is the safety of people and property. Fly in a large field, preferably without a crowd present. If you do have spectators, make sure that they are aware that you are flying an untested rocket and that they stay a good distance away from the rocket when you launch it.

MAKING YOUR OWN COMPONENTS

Once you have built and flown several kits and some designs of your own, you may wish to turn your hand to making your own components, rather than buying them from the manufacturers.

Body Tubes

Body tubes can be made from gummed *kraft tape*, of the kind used to seal shipping cartons. You will need a wooden dowel of the same diameter as the inside of the tube you wish to make, around which you will wrap the tape.

The first step in making a body tube is to cut a length of the tape. (The length will depend upon the length of the tube.) Both ends of the tape should be trimmed to a 45-degree angle, and the tape laid flat on the surface of the work area with the gummed side facing up. With a wet sponge or cloth, wet a strip approximately 1/4-inch wide along one side, as in Fig. 3-26.

Next, lay the rod on the top, dry corner of the tape, as in Fig. 3-26, and begin wrapping the tape around the rod, taking care not to wrap the tape unevenly. An uneven wrap can result in the tape sticking to the rod and ruining the tube.

Once you have wrapped the tape completely around the rod from end to end, set it aside to dry. Cut another length of tape, equal to the first, and trim the edges to a 45-degree angle. Moisten this strip completely and wrap it around the first, as in Fig. 3-27.

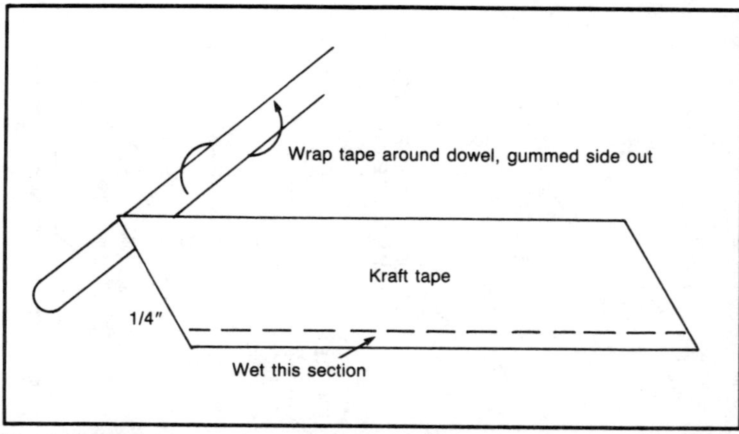

Fig. 3-26. Preparing kraft tape for wrapping a body tube.

Slide the dowel rod out of the tube and set the tube aside to dry; it is complete.

Nose Cones

Nose cones can also be made from kraft tape (Fig. 3-28). Cut one end of a length of the tape to a 45-degree angle and lay it flat, gummed side up. Wrap point A to point B in the direction shown in Fig. 3-28, so that point C becomes the tip of the nose cone. Following this, moisten the remaining length of tape and wrap it tightly around the cone.

After the nose cone has dried, you should make a small cylinder of kraft tape, equal in diameter to the inner diameter of the body tube that you plan to use with the nose cone, and glue it inside the base of the nose cone. This will hold the nose cone in the body tube (Fig. 3-29).

If you enjoy working with wood, you will find balsa nose cones (or nose cones made with other types of wood) easy to make. Starting with a balsa block, first cut it into the rough shape of a nose cone. After this, drill a small hole into the base of the block and glue a dowel rod in the hole. Place the opposite end of the dowel rod in the chuck of a drill, and begin turning the block at slow speed. Using a razor knife, "turn" the rough nose cone into the finished product, just as you would on a lathe.

Recovery Devices

You will notice that the parachutes packed with most model

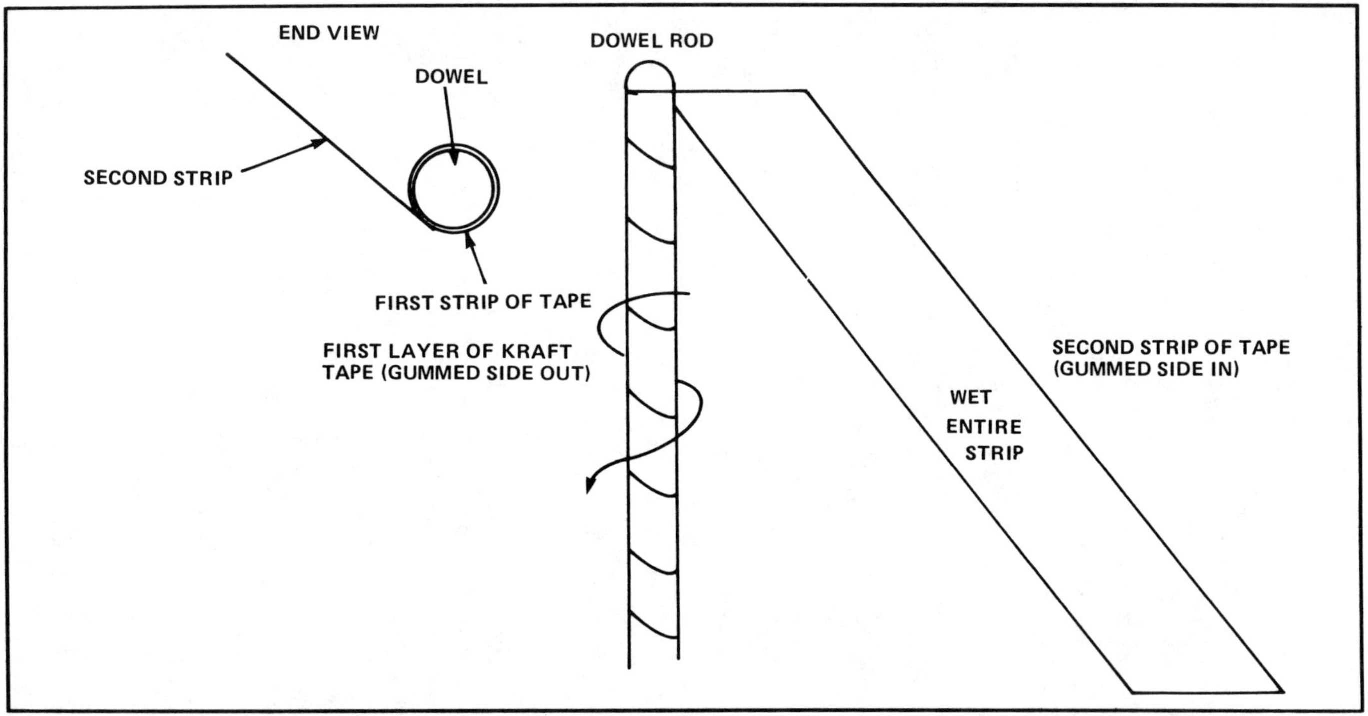

Fig. 3-27. Wrapping second layer of tape for body tube.

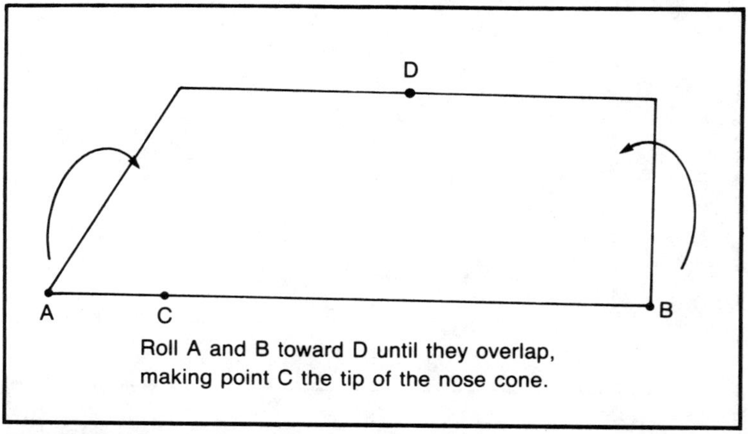

Fig. 3-28. Wrapping kraft tape into a nose cone.

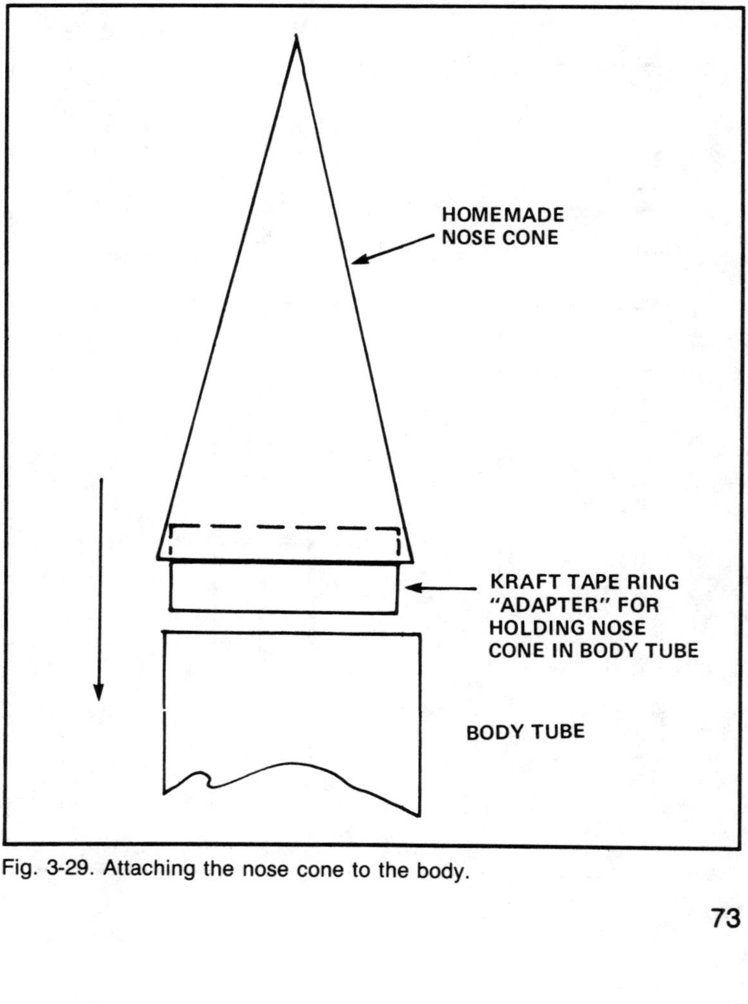

Fig. 3-29. Attaching the nose cone to the body.

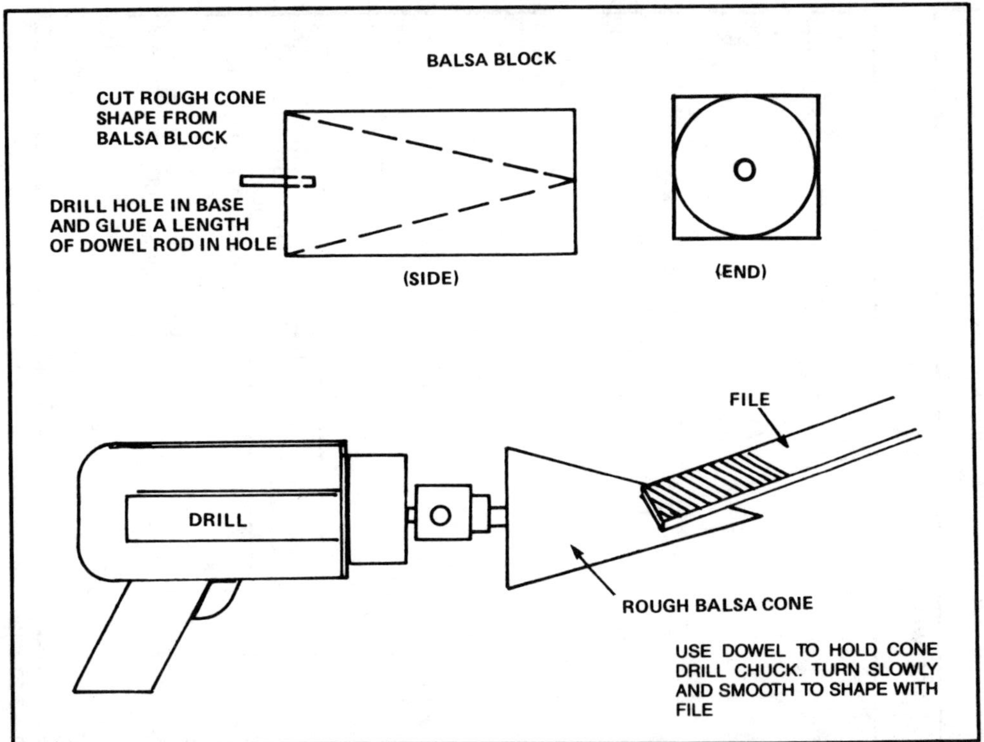

Fig. 3-30. A nose cone can be shaped from a block of balsa.

rocket kits (and those sold separately) are made from lightweight plastic, usually brightly colored and bearing the manufacturer's logo. You can make your own parachutes, assembling them in the same manner as those provided by the manufacturers, using plastic garbage bags and kite string (Fig. 3-31). Garbage bags are even stronger than the plastic the manufacturers use, and you may wish to use heavier string, also. If you use garbage bags, use black ones, because black is easier to see against a blue or gray sky.

If you wish to construct lighter parachutes, silvered mylar plastic is both lightweight and easy to see.

Streamer material provided by most manufacturers is flameproof paper; crepe paper is the same, and is excellent streamer material. Again, dark colors are highly visible, and silver mylar offers both light weight and high visibility.

ENGINE MOUNTS

Engine mounting tubes can be made with kraft tape, in the same fashion as body tubes. (A flameproofing solution is recommended.) Engine clips can be fashioned from strips of aluminum.

Another more difficult but more professional technique for fabricating model rocket body tubes is presented in William S. Roe's technical report, *Rolling Body Tubes*, published as National Association of Rocketry (NAR) Technical Report #10.

(This report is reproduced here by the permission of the National Association of Rocketry, 182 Madison Drive, Elizabeth, PA 15037.)

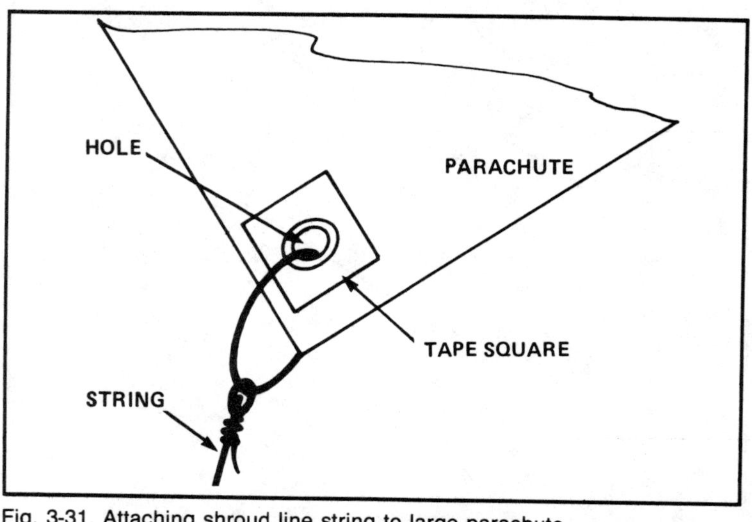

Fig. 3-31. Attaching shroud line string to large parachute.

ROLLING BODY TUBES

by William S. Roe, NAR #13

"The rolling of a paper body tube by hand for use as an air frame for a model rocket is truly an art. For those needing great quantities of tubes, such as would be needed by an NAR section, the art can be transferred to a machine. By use of a machine, 25 to 30 tubes can be rolled in an hour. They will be uniform in size, exactly the same weight, and will be tightly rolled. If desired, the ends can be trimmed on the machine so as to be absolutely square.

"There are probably several ways to make a machine suitable for tube rolling. One factor which would dictate the kind of machine desired is the rate of production required. Another factor would be whether such a machine must be purchased or whether a machine on hand could be converted. The tube rolling attachment described in this report was designed to fit a Model 10ER Shopsmith Combination Tool. The design would fit equally well on the newer Mark V Shopsmith. Quite possibly, after reading this report, you will find that a wood lathe of some description can be converted to a tube roller.

"Basic to any tube rolling machine is the *shaft* or *mandrel* around which the tube is actually rolled (Fig. 3-32). The mandrel

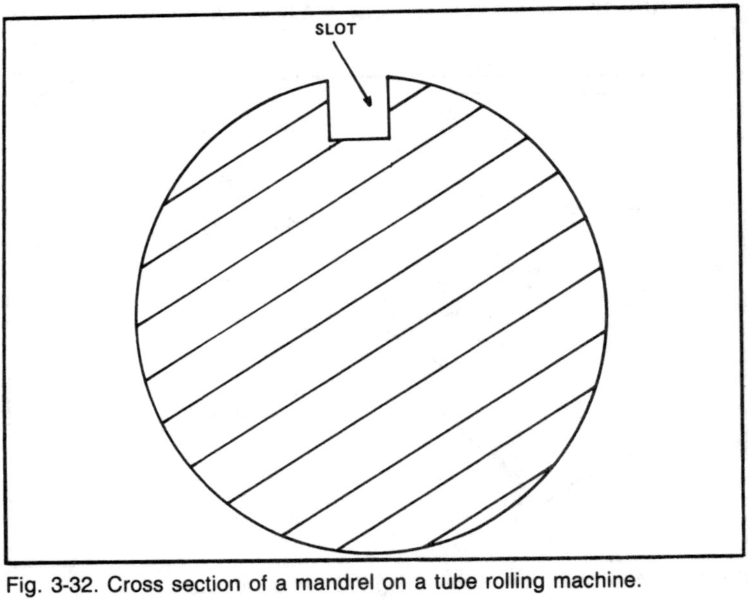

Fig. 3-32. Cross section of a mandrel on a tube rolling machine.

should have a diameter equal to the desired inside diameter of the tube to be rolled. A slot must be machined or milled along the entire length, as shown in Fig. 3-32. This slot should be square, 1/32 inch wide and 1/32 inch deep. A wider slot will make it difficult to keep the paper in, as will a shallower slot. A deeper slot will leave more of an unglued tab on the inside of the tube. While this unglued tab is not harmful, it makes it more difficult to fit nose cones into the tube. One end of the mandrel shaft is turned down to a diameter that will fit the Jacobs chuck, which is standard equipment on a Shopsmith. A 1/2-inch diameter and 1/2-inch length on the turned down end is usually sufficient to give the chuck a good grip on the mandrel.

"The other end of the shaft should be turned to a size to fit whatever ball bearing is available, or should be machined out in a cone tapered shape to match a tailstock live center. These two devices will be discussed later. The overall length of the mandrel shaft should be 22 inches. This will allow tubes up to 18 inches long to be rolled. If the mandrel is made shorter, corresponding reductions will also have to be made in the bed plate and base board described later. The shaft should be made absolutely straight and should be made of a good grade of steel, although it need not be hardened.

"The bed plate against which the rolling is done is a piece of 3/4-inch board 11 1/2 inches wide and 21 inches long. It is covered on one side with a sheet of stainless steel turned down around the edges of the bed plate and fastened with screws so that the surface is not blemished. A sheet metal shop should have no trouble covering this board for you.

"The bed plate is hinged to a base board (Fig. 3-33), which is another piece of 4/5-inch board of the same size as the base plate. Any type of 3-inch hinge is suitable. The hinges are positioned so that when pulled forward the bed plate is at right angles to the base

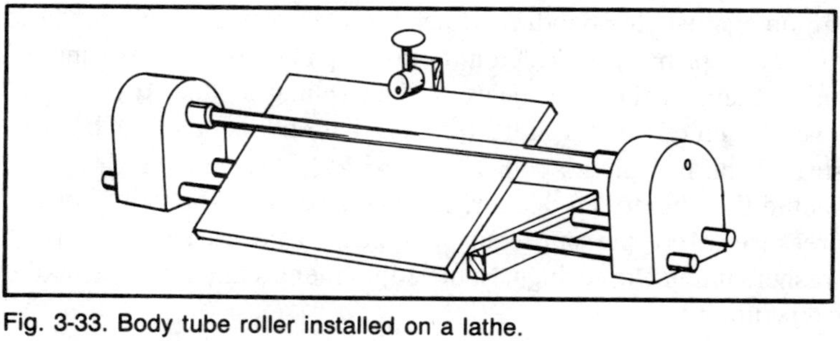

Fig. 3-33. Body tube roller installed on a lathe.

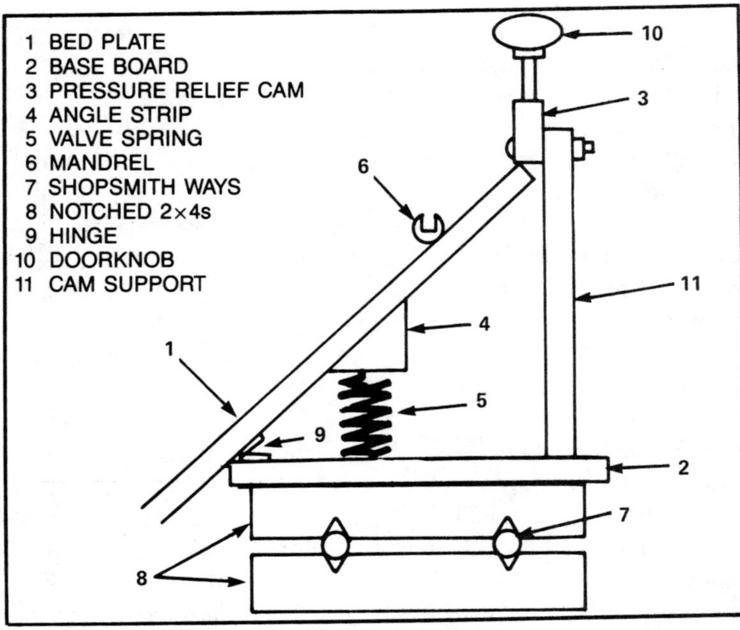

Fig. 3-34. End view of tube rolling attachment.

board (Fig. 3-34). The top of the bed plate should be positioned so that the mandrel is parallel to the long end of the bed plate and about 3 inches from the top of the bed plate.

"The base board is fastened to the ways of the Shopsmith with two notched 2 × 4s cut fit the ways. Another 2 × 4 is notched the same way and put under the ways. A pair of long carriage bolts are used to make the 2 × 4s clamp the ways and hold the whole assembly tight.

"Pressure is applied between the mandrel and the bed plate by two automobile valve springs. The bottom of the bed plate is equipped with a strip of wood cut at an angle so that the bottom of the strip is parallel to the base board when the bed plate is resting up against the mandrel. Holes large enough to accommodate the valve springs are drilled into the strip and also directly underneath them in the base board. When rolling a tube, if the tube wrinkles on one end it indicates that that end has too much pressure of the bed plate against the mandrel. The hole on that side should then be drilled deeper, to relieve some of the valve spring pressure. More pressure can be applied if necessary by putting washers under the springs. Some adjustment will have to be made from time to time.

"You must be able to move the bed plate away from the mandrel to insert the paper at the start of the rolling operation and to remove the tube and the mandrel at the end of the rolling operation. Therefore, some method must be used to move the bed plate away and hold it away as desired. The pressure is relieved by a circle of wood mounted off-center like a cam. It is placed on a vertical support, behind the bed plate, so that it pushes the bed plate down when turned. Not much relief is required. The edge of the cam should be covered with sheet metal to make it wear better, and a handle, such as a doorknob and shaft, should be added to the cam for ease in turning it.

"The tailstock piece of the Shopsmith is used. The best fitting to accommodate the mandrel is a tailstock live center, available as a Shopsmith attachment. This center is tapered at an angle of 60 degrees. The end of the mandrel should be machined out to match. An adaptor piece can be made which will do the job of a live center, but it requires expensive machining. One end of the adaptor piece must be made to fit the Shopsmith tailstock chuck arbor (standard accessory) with a ball bearing on the other end. The inner race of the bearing is used to support the mandrel.

"Begin tube rolling operations by having the quill on the headstock fully retracted. Put the mandrel into the Jacobs chuck attached to the headstock and tighten the chuck with the key. Then extend the quill while guiding the other end of the mandrel into the tailstock fitting. Lock the extension into position. While pressure is being relieved between the mandrel and the bed plate by the cam, insert the edge of the paper into the mandrel slot. While holding the paper in the slot with one hand, apply pressure from the bed plate by turning the relief cam. The paper must stay in the slot. Then, by hand, turn the shaft towards you until the paper just meets itself around the mandrel.

"The machine is now set for the actual rolling. Apply glue with a 3-inch brush to the exposed paper on the bed plate. Do not use too much because the excess will be squeezed out along the seam of the tube, making a mess and an unsightly tube. Be sure that the edges of the sheet have glue, however, especially the lower edge, since it forms the outer seam and must be stuck tightly.

"The lowest speed on the Shopsmith should be used. It is necessary to apply power only for the flick of a switch. This will make the tube roll about ten times. More rolling than this will simply wear off paper, making a 'fuzzy' tube. Roll the mandrel by hand until the seam is directly between the mandrel and the bed plate, where the greatest pressure is being applied to it. Leave it there while you use a damp rag to wipe the excess glue off the stainless steel bed plate. Then, relieve the pressure with the cam and turn the mandrel to make sure the outer seam on the tube is stuck tight. If it is not, turn it until pressure is on it again and wait. Excess glue can be wiped off the tube at this point, or extra glue can be added to any spots that may have been missed. Any smooth rod, such as a knitting needle, can be used to further smooth the seam.

"At this point, the tube can be trimmed on the ends if necessary, or it can be cut to the desired length. A heavy, sharp knife is the best tool. With the pressure off and the machine turning at low speed, apply the knife at right angles to the tube and hold firmly. If you do not hold the knife firmly, it will thread itself down the tube and all is lost!

"The tube can now be removed. Take the mandrel out of the chuck and off the machine. Slide the tube off the mandrel. It should come off easily. If it does not, you probably did not turn the mandrel far enough at the beginning so that glue got onto the shaft. Remember that none of the shaft should be visible when glue is applied.

"The use of a knife to cut tubes will mar the mandrel somewhat, and a file should be run along it while it is turning to smooth this out. The use of floor wax on the shaft may be of help, but is not necessary.

Paper to Use

"The best material for tubes is not paper, but *tagboard*. Since it is made from the same materials and in much the same way as paper, it is simply a heavier paper. It is much tougher because of special materials and different treatment. It is called 'tag' because it has certain uses and denotes the material itself. The designation means a great deal to those who work with paper.

"Designations for the hundreds of different kinds of paper on the market are complicated. Follow the specifications given below to be certain that you are getting the right kind of tagboard for rolling tubes. A print shop or stationery store next to a print shop is the best place to purchase tag stock; the people at these places will also understand the specifications. For a 14-inch tube with 3/4-inch inner diameter (ID), ask for stock cut to 9 1/2 × 14 inches from 100-pound basic tag stock. (This means that 500 sheets, or one ream, of this paper in a base size of 24 × 36 inches will weigh just 100 pounds.) This paper is obtainable in white, but 'manila' (a light tan color) is more common, and may be all that you can find. Its working properties are the same and paint will cover it just as well. A tagboard with a base weight of 125 pounds can be used to roll tubes, but it is awkward, harder to work, and makes a heavier tube. Tag also comes in 150-pound weight, but is just too heavy to work at all.

"For a 14-inch tube with 3/4-inch ID, start with a piece of paper 9 1/2 × 14 inches with the 'grain' parallel to the 14-inch dimension. This is very important. If the grain is wrong, it will be impossible to roll a tube. All paper has a grain parallel to the minute fibers of the wood from which the paper was made. The paper will fold or roll smoothly only if the fold or roll is made parallel to this grain. (Although it is of no concern for rolling tubes, it is interesting to note that when paper must be folded across the grain, it is scored by machinery to break the fibers and enable the fold to be made.)

"For rolling tubes of larger internal diameters, a larger mandrel is needed and different paper sizes are required. Other tube sizes, such as 7/8-inch and 1-inch ID, can be rolled using larger mandrels. It may be necessary to adjust the pressure relief cam for these larger sizes. Paper stock for 7/8-inch and 1-inch tubes 14 inches long should be cut to 9 11/16 × 14; in fact, those who wish to use a standard size paper to roll tubes of different sizes would do well to standardize on the larger paper size, since it also rolls 3/4-inch ID tubes well.

"Another kind of paper may also be used to roll good tubes. It is readily available from print shops, and comes in a variety of colors. Ask for *folding bristol*, 125-pound. This weight is also known as '2-ply.' Be sure it is cut with the grain running parallel to the long dimension.

"With this information, and a little arithmetic, you can also roll tubes of bigger and smaller diameters, different thicknesses, and different lengths up to 18 inches on the machine."

Glues to Use

"Almost any type of glue made for paper will work, even flour and water paste. If you are trying for speed of production, however, there are special glues made for box makers and bookbinders that are designed to stick; then dry rapidly when applied, yet can be worked before they set.

"An excellent white glue, used by bookbinders, is 'YES,' made by Gane Brothers & Lane, with offices in New York City, Chicago,

San Francisco, and St. Louis. This glue comes in about the consistency of soft butter, and should be thinned with water in the proportions of two parts by volume of glue to one part of water. It can then be brushed readily. Incidentally, this is a good all-around glue to use with paper. It dries flat, without wrinkling, and any excess can be removed with water.

"Two companies make an excellent dry powder glue as a derivative of corn starch. Both are made into usable liquid in the same manner. They are easy to work, keep well, have superior sticking qualities, and any excess can be removed from tubes or machinery with a damp rag.

"The National Starch & Chemical Company, 750 Third Avenue, New York, NY 10017, makes 'Aqua Flakes #15.' The Corn Products Company, Argo, IL, makes 'Dextrine.' Both of these products must be mixed and cooked before being used. Mix thoroughly one cup of water to one and a half cups of glue. Heat the mixture almost to the boiling point and allow to cool. After cooling, the graininess will have disappeared, and the glue is ready to use. It should be used cold. 'Note: Don't try paper mucilage; it wrinkles the paper.' "

Conclusion

"If you have tried to roll tubes by hand, the information herein will help you roll better ones. And, if you are tired of laboriously rolling tubes by hand, you will really appreciate the machine described in this report, which is capable of producing hundreds of tubes all alike. Be sure to wipe off all excess glue as you go along; it pays to keep the machinery clean and free of dried glue.

"After you have completely mastered the art and learned how to adjust the machine, you will probably want to experiment. A thinner walled body tube, more difficult to use but lighter, is made from paper 5 × 14 inches. You can also try rolling a tube with one layer of aluminum foil for a liner. Get a painter or printer to give you some fluorescent paper or cardboard; it is tricky to handle because any glue coming out along the seam will remove the colored coating. The tubes made with this paper will be of heavier material, but won't need painting."

As with all model rocketry activities, the fabrication, machine operation, mixing, and other activities described above should be undertaken will all possible safety equipment in place, and with utmost caution.

Chapter 4

Flying Your Model Rocket: Launch Range Requirements and Practices

Model rocket flights cannot be made from the average back yard or vacant lot. The tendency for a rocket to drift horizontally during the recovery phase makes it hazardous to fly in areas where there are large numbers of trees and/or buildings. If you fly in small fields, you will lose many of your rockets to trees or roofs, and you may do some property damage.

If you confine your flying to A-, B-, and C-powered models, a school football field or baseball diamond will probably provide sufficient recovery area, provided there is little or no wind. For larger models, or for low-power rockets built for high altitude performance, you'll want to look for something larger. Flying boost-gliders and rocket-gliders will also call for larger fields.

SELECTING A LAUNCH SITE

In general, the *shortest* side of your launch field should be equal to at least one quarter of the maximum altitude your rockets will reach (Fig. 4-1). Hopefully, the area surrounding your launch site will be fairly clear of trees and houses as well, in case a stray wind sends your rocket out of the recovery area.

In addition to being roomy and as far as possible from houses and trees, your launch site should be cleared of any easy-to-burn materials and located away from power lines. You should also determine whether or not the site is under the approach path to any nearby airports. If it is, try to find another launch site or at least

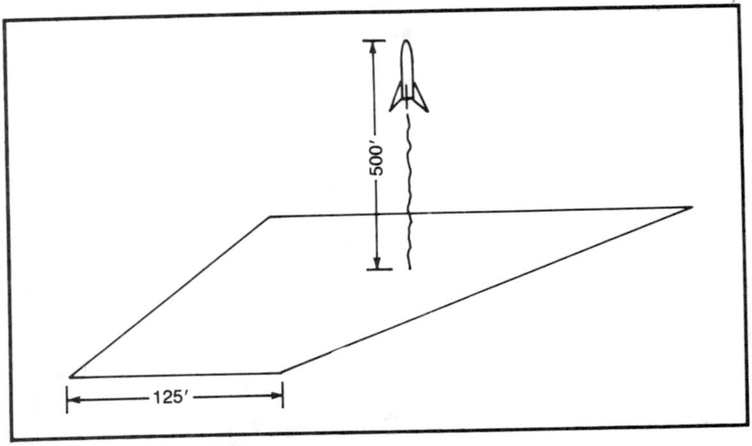

Fig. 4-1. Minimum launch field dimensions.

fly only during periods when you know there is no air traffic—if this is possible.

If you cannot find a field suited to your requirements, think about scaling down your flight activities. Flying low-power model rockets is better than not flying at all!

Model rocketry is not a particularly noisy activity, so you won't have to worry about disturbing nearby residents when you fly. You should, however, avoid flying when there are high winds (over 15 mph), as you will risk having a rocket land on someone's roof (not a good way to make friends!). You should also avoid flying at night, unless you are experimenting with, for example, a strobe light as a payload. Night flying not only reduces your chances of recovering a rocket, but also may draw unwanted attention to your activities because the exhaust of a model rocket engine can easily be confused with fireworks by those not familiar with model rocketry.

Be sure that you have permission to use a field as a launch range before you fly, even if the field is a public park. Model rockets can be confused with fireworks and moving into a field without permission will usually get you thrown out—permanently.

The best way to get permission is to set up an appointment with the field's owner (or the person responsible for a public field) and explain to that person exactly what you will be doing. Take along a rocket or two, along with explanatory literature (such as catalogs or this book), and make it clear that you are not shooting off fireworks. Stress the excellent safety record of model rocketry; it is far better than that of model airplane flying or even baseball. Let the person responsible for the field know, too, that you won't be making a mess of things and that you will be flying in a manner that will present no hazards to other people or property.

You will find that nearly anyone you approach will be agreeable to your model rocketry activities (Fig. 4-2), but some owners of potential launch sites may be concerned about liability insurance. If you find this to be an obstacle, membership in the National Association of Rocketry can be of great help. The Association makes available to its members a million dollar liability policy for less than $10 per year, and the coverage can also be extended to the owner of a launch site.

EQUIPMENT AND SUPPLIES

In addition to the rockets you want to fly, engines, igniters,

Fig. 4-2. A model rocket on launch pad.

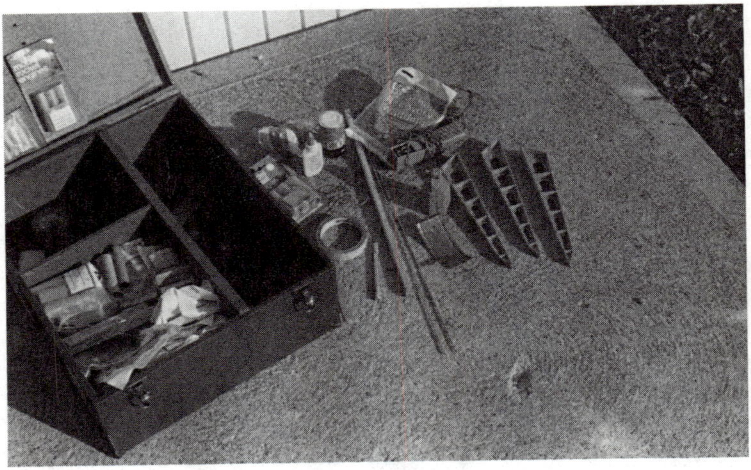

Fig. 4-3. Typical range box.

recovery wadding, and a launch system, you will want to put together a *range box* (Fig. 4-3) to take with you whenever you fly.

A typical range box can be a tool box or tackle box, and it should contain the following items: spare parachutes and streamers, spare shock cords, razor knife, needle-nose pliers, talcum powder, quick-setting epoxy, masking tape, extra batteries, electrical tape, shroud line, scissors, sandpaper, and spare micro-clips for launch controller.

You'll also find the range box a handy place to keep your engines, igniters, and recovery wadding. The items on this list are for any emergencies (such as a shredded parachute) you may encounter while you are flying and serve as "flight insurance" against loss of flying time. After you've been flying for a while, you will undoubtedly think of other items to help with field repairs, but this is the minimum you should carry.

Safety equipment is simple. A *fire extinguisher* is highly recommended, particularly if you are flying in dry, grassy areas. If you can't buy one, a reliable extinguisher can be made using a common pump-powered garden sprayer filled with water. In addition to an extinguisher, you may want to practice a little fire prevention by placing a tarp under your launcher to minimize the chance of hot sparks from an engine's exhaust touching off a grass fire. Finally, a small first-aid kit may be desirable.

To help track your rockets, particularly the high-power models, you may want to take along a pair of binoculars or a telescope. In addition, a can of powdered *tempra paint* will prove invaluable in helping you spot your model rockets at high altitudes. The paint should be poured on top of a model rocket's parachute or streamer before you place the nose cone on the body tube. At ejection, the powder will be blown out ahead of the parachute and scattered, making a small, highly visible cloud. Blue is recommended for clear days, and red is recommended for overcast days. Black is good under any conditions.

LAUNCHER LOCATION

If you are flying with no wind at all, the ideal location for your launch is in the center of the launch field. Windless days are rare, however, so you will usually have to make allowances for wind in locating your launcher.

Because model rockets tend to weathercock, or lean into, the wind, you will find most of your rockets returning to you from the direction the wind is blowing. Weathercocking can be compensated for by tilting the launch rod (Fig. 4-4) away from the wind (though

Fig. 4-4. Tilting the launch rod to prevent weathercocking.

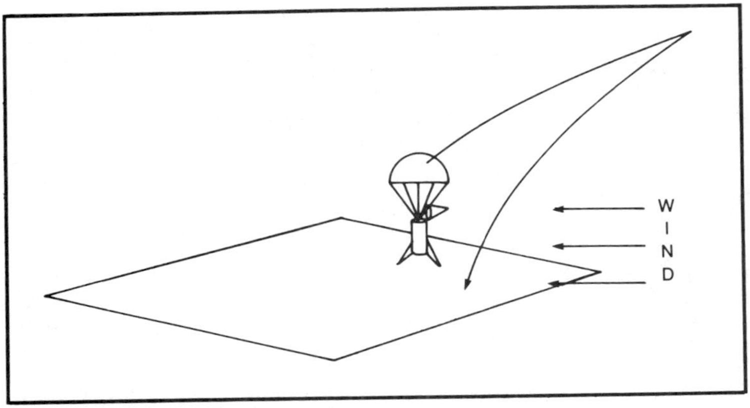

Fig. 4-5. Launching into the wind.

you should never tilt the launch rod more then 30 degrees away from vertical), but this usually is not enough to prevent the rocket's heading into the wind somewhat.

To overcome recovery problems created by weathercocking, you should place your launcher on the side of the launch field nearest the direction from which the wind is coming, as in Fig. 4-5. This way, your rocket will be blown back toward you after it flies into the wind, and the entire length of your field will be available for recovery.

If you don't want to fly from the edge of your field, you can tilt your launch rod slightly into the wind so that the rocket's flight will carry it upwind of your field (Fig. 4-5). This will result in the rocket being blown back into the recovery area by the wind.

PREPPING THE ROCKET

Flight prepping (Fig. 4-6) a model rocket is a relatively simple operation, but it is easy to foul up, especially if you are in a hurry. For this reason, you should take extra care during preflight preparations. It only takes one mistake to cause flight failure, which means a wasted engine and igniter, not to mention the possibility for damage to the rocket or property.

Normal preflight prepping includes the following:

1. Install igniter in engine, taping it in place (Figs. 4-7 and 4-8).
2. Install engine in engine mount.
3. Place recovery wadding in forward end of body tube.
4. Fold parachute or streamer (Fig. 4-9), adding talcum

Fig. 4-6. Rocket prepped on the launch pad.

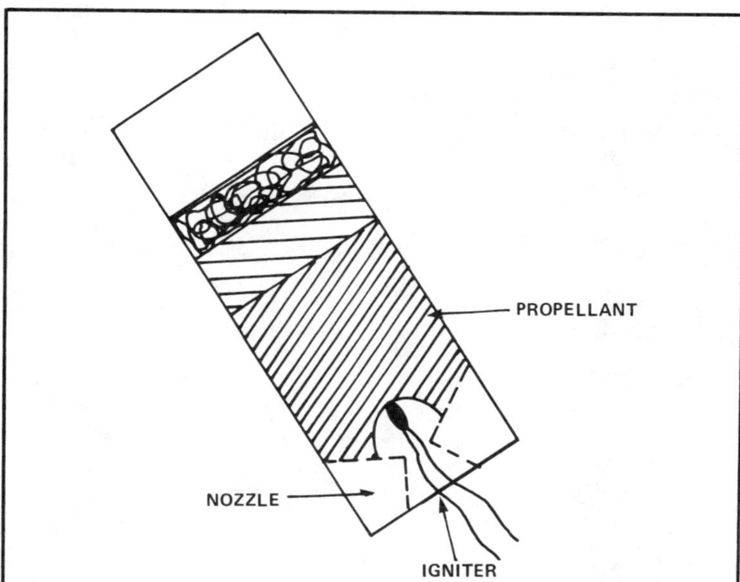

Fig. 4-7. Proper igniter installation.

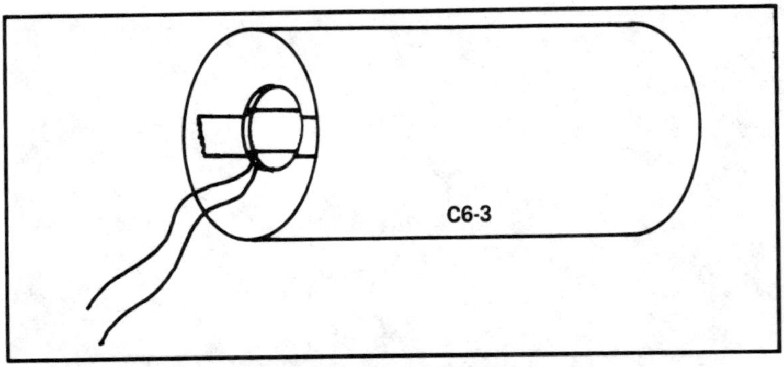

Fig. 4-8. For preflight preparation, tape the igniter in place.

powder to aid chute opening, if desired.

5. Wrap shroud lines and shock cord and insert in forward end of body tube with parachute or streamer.
6. Put nose cone snugly in place (Fig. 4-10).
7. Check fins, launch lug, and other rocket components for possible damage.

Now your rocket is ready to fly. Once it is prepped, you slide

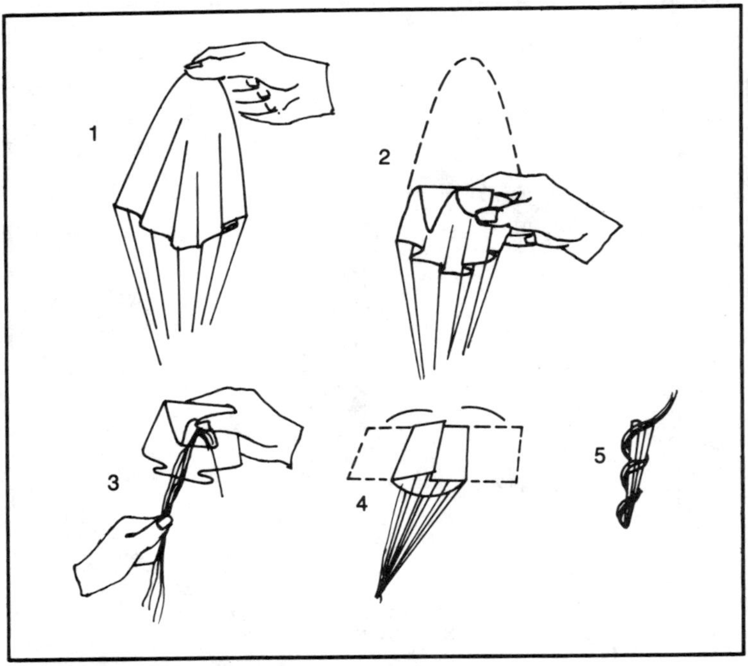

Fig. 4-9. Folding a parachute.

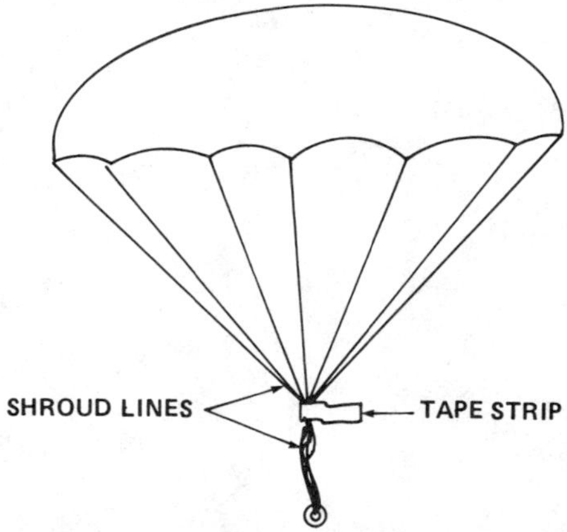

Fig. 4-10. Reefing a parachute by taping shroud lines.

it over the launch rod (rod through the launch lug), attach the micro-clips to the igniter wires, and give everything a final visual inspection.

This completed, you are ready for launch. You should first check the launch area to be certain that no people or animals are heading for the launcher, and check the sky for low-flying aircraft.

If everything is clear, you should insert your safety key, check for continuity by making certain that the continuity light is on (this means that you have good connections throughout the launch system) and begin a brief countdown, counting backwards from five. When you reach one, press the launch button (Fig. 4-11). The continuity light should go out, and the engine ignite. (After the rocket has lifted off, be sure to remove the safety key from the launch controller. There is nothing more surprising than connecting micro-clips to igniter leads and having the engine ignite under your nose! This can happen if the launch button happens to stick and you've left the safety key in the controller.)

In case of a *misfire*—meaning the rocket doesn't ignite— wait at least one minute before approaching the rocket. The engine could very well ignite suddenly, and you don't want to be standing over the rocket if this happens.

Check for possible causes for the misfire carefully. If the igniter didn't ignite, you may have old or weak batteries in your launcher. If the batteries are okay, you may have attached the micro-clips to the igniter leads improperly—they may be touching one another, or touching the blast deflector. You may also find that

Fig. 4-11. Liftoff! Courtesy Estes Industries.

the micro-clips have a build-up of dirt on them (sometimes a by-product of the exhaust of rocket engines), in which case they need to be lightly sanded. You may also have a bad igniter.

If the igniter did burn, the problem may be that you installed it incorrectly. Perhaps it wasn't fully inserted in the nozzle (it should be pushed in until its tip touches the propellant inside the engine). Or, you may not have taped the igniter securely enough, which permitted it to fall away from the engine.

If all of these items check out, and you've attempted to ignite the engine several times, you may have a bad engine. If you suspect this to be the case, return the engine to the manufacturer, along with a note explaining the problem. You will most likely receive a replacement engine.

(On the subject of defective engines, you should be aware that model rockets should not be stored in hot car trunks, metal boxes, or other places where the temperature is likely to exceed 160° Fahrenheit. This temperature is a critical one; model rocket engines that have been exposed to higher temperatures can *blow through*, which means that all of the propellant, as well as the ejection charge, can ignite at the same time. This can result in a destroyed or damaged rocket and launcher.

If you should experience such a catastrophic failure with an engine that you know was not exposed to high temperatures, return what is left of your rocket and the engine casing to the manufacturer for replacement, along with any other unused engines that were packaged with the defective engine.)

RECOVERY

A successful recovery is as important as a successful flight to model rocketeers, because you'll want to fly your model rocket more than once and may need to recover an important payload. Naturally, you'll want to maximize the chances of successful recovery.

You should, of course, select the best recovery system for both the size and type of rocket you are flying and the expected performance and flying conditions. A small, lightweight rocket will drift a great distance if you use a parachute for recovery, so a streamer or tumble recovery is the best technique to use. A large, heavy rocket will not drift far, but high winds may carry it farther than you expect, so it is a good idea to *reef* the parachute shroud lines (tape them together about halfway between the parachute and the point where they attach to the nose cone) and maybe even use a smaller chute in high winds. Rockets that fly to very high altitudes will have more opportunity to drift, because their descent time is greater and they may encounter winds at high altitudes that are not detectable at ground level. You may find that a streamer is the best recovery system for rockets that can achieve high altitudes.

These are just a few examples of the various aspects of performance, size, and flying conditions you will have to take into consideration. Weigh all factors carefully before choosing your recovery system, and you will greatly enhance your rocket's recoverability.

As discussed earlier in this chapter, the location of your launcher is very important in determining whether or not your rocket will return to the recovery area. *Tracking* your rocket helps a good deal during recovery (Fig. 4-12). The use of powdered tempra paint, as mentioned earlier, is highly recommended, because it is very important to be able to see just where the rocket is at parachute ejection. You should have some help in tracking the rocket, too. A recovery crew of at least two people should be out on the launch range a good distance from the launcher when the rocket is launched. The farther they are, the better, because it is sometimes easier

Fig. 4-12. Tracking phase of rocket's flight.

to observe the complete flight path of a model rocket from a distance than it is from beneath. Thus, if you loose sight of your rocket during its ascent, all will not be lost, because at least one of your range people should be able to follow it.

Once the rocket's recovery system has deployed and the rocket has descended a bit, it should be easy to spot by anyone on the launch range. If the rocket flies away from the field during flight, though, it will be important to have those observers situated at a distance.

Hopefully, all of your rockets will end up in the designated recovery area. If one doesn't, don't panic. Try to keep it in sight as it drifts, and, when it disappears behind trees or buildings or over the horizon, try to spot a large object (such as a tree) that will serve as a landmark to indicate where the rocket touched down. Sight on this landmark and walk a straight line toward it to the area in which you believe you will find the rocket. Allow a little extra distance, though, as the rocket usually appears to be closer to you than it really is.

Search the area carefully, with help if possible. Surveying an area from a high vantage point (a tree or an upper floor of a building) can often be of great help. If you do lose a rocket, try to analyze just what went wrong during the flight or descent phase and plan to avoid that problem during future flights.

Chapter 5

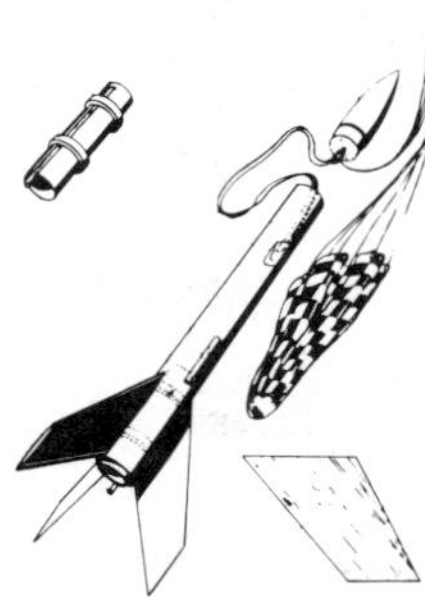

Beyond the Basics: Gliders and More

Once your level of experience has advanced beyond that required for building and flying simple single-stage models, you will want to take on new challenges. This chapter will cover some intermediate-stage model rocketry activities.

As with any and all model rocketry activities, you should keep safety first in your mind when undertaking projects of the types discussed in this chapter. Safety in model rocketry cannot be overstressed; the fact that model rocketeers are so safety conscious is one reason for the hobby's enviable safety record. *Remember practice safety for its own sake—and yours.*

ROCKET-GLIDERS

A rocket-glider is a model rocket that is recovered using a glide technique. The biggest difference between a rocket-glider (better known as R/G) and a boost-glider (discussed later in this chapter) is the fact that the R/G carries back to earth with it everything that it carried on liftoff (Fig. 5-1). The engine is not jettisoned, as is the case with boost-gliders.

Most R/Gs have the appearance of model airplane gliders (Fig. 5-2). This, of course, is the most stable configuration for the glide back to earth, as well as being the most effective design for competition. The main differences between R/Gs and model airplane gliders are that R/Gs have an engine mounted on them and that

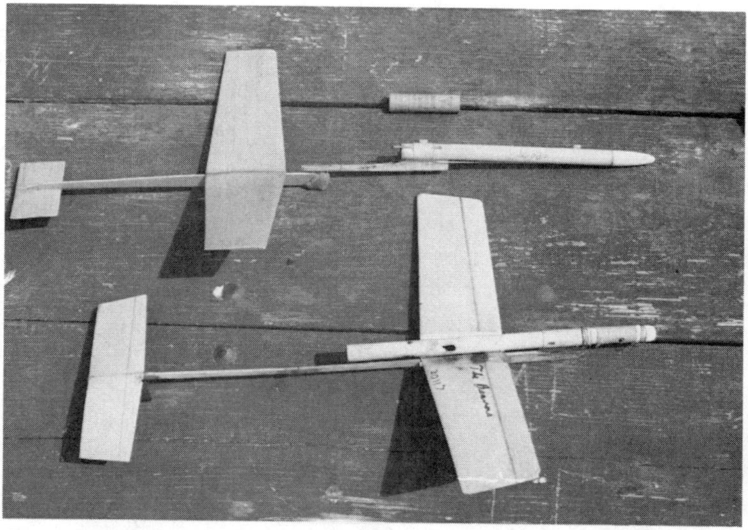

Fig. 5-1. Boost-glider (top) and rocket-glider (bottom). Courtesy Tom Kolis.

the rudder is usually mounted beneath the fuselage, instead of on top, as with a normal aircraft. The rudder is mounted this way to protect it from the engine exhaust, and it does not affect the stability of the glider in any negative way. The rudder does help stabilize the R/G during its boost phase, though, as the cross section of an

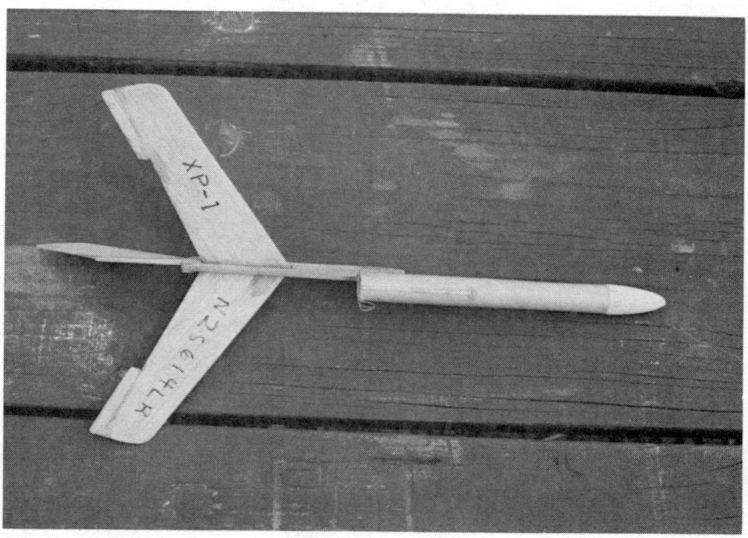

Fig. 5-2. Rocket-glider. Courtesy Tom Kolis.

R/G shown in Fig. 5-3 illustrates. As you can see, the rudder *offsets*, or balances, the drag created by the engine pod. It creates, with the wings, an aerodynamic effect similar to that created by the fins of a four-fin rocket.

Looking at a model aircraft in terms of stability requirements of a model rocket, it would appear that the aircraft (in this case a glider) would be highly unstable, the forward cross section just discussed aside. The wings function as large forward fins, which tend to move the CP forward (a definite *no* for stability, as discussed in Chapter 4). The stabilizer is too small to offset this effect.

How, then, can the aircraft configuration be stabilized as an R/G? Simple: as discussed in Chapter 3, the CG must be ahead of the CP, so you move the CG forward. This is accomplished by adding weight to the nose. Because adding weight to the nose can be accomplished by lengthening the fuselage, you move the CG forward by adding a long boom and mounting the engine on that boom. Figure 5-4 illustrates the typical configuration of an R/G and shows how the boom-mounted engine adds weight to the forward end of the model.

If you know anything about the stability of aircraft, you know that a lot of weight in the nose of an aircraft will ruin its stability. This cannot be compensated for by constant readjustment of the control surfaces in a model, however. You are left with a model that flies with good stability as a rocket, but that would, if left unchanged, dive straight to the ground as soon as the boost-phase of the flight ended.

Fortunately, there are techniques for changing the stability configuration of an R/G from that of rocket to that of glider during

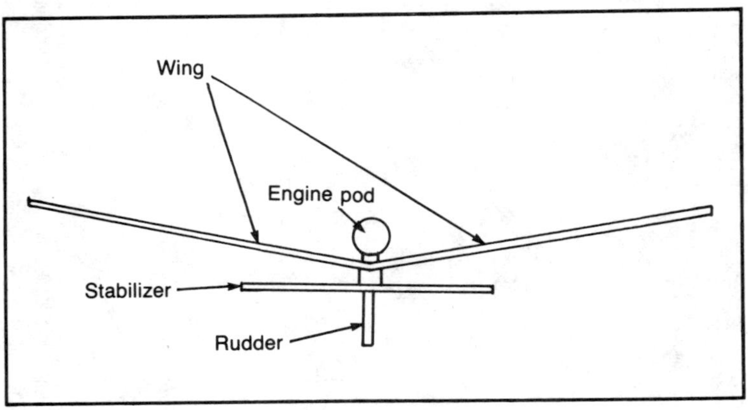

Fig. 5-3. Forward cross section of a rocket-glider.

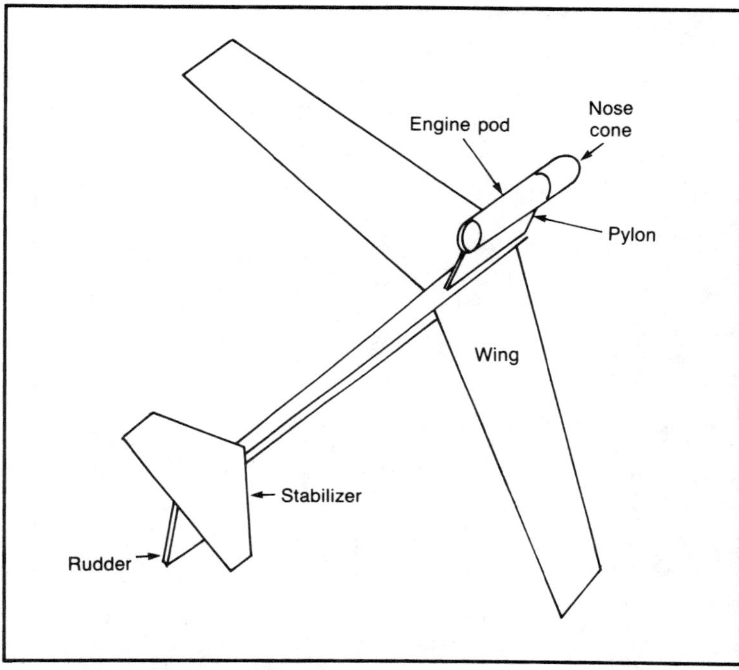

Fig. 5-4. Rocket-glider configuration.

the flight. Once the boost phase has ended, it is necessary to move the CG toward the rear, which in turn causes the CP to move forward, ideally resulting in a stable glider.

The most often used technique for accomplishing this change is to move the engine rearward at ejection. As shown in Fig. 5-5, this is accomplished by constructing the engine mounting tube in such a way that the engine is at the forward end of the mounting tube during boost and is kicked toward the rear by the force of the ejection charge.

It is also possible to use the ejection charge to change the *trim* (fin position) of the model. This can be accomplished by having the engine move backward against a lever that causes the flaps on the stabilizer to move up, thus forcing the nose up. It is also possible to set up a mechanism whereby the ejection charge burns through a cord that holds the stabilizer flaps horizontal. At burn-through, a rubber band then pulls the flaps into their "up" position. Figure 5-6 illustrates both of these techniques.

If you are planning to fly R/Gs, it is best to seek out R/G kits from the manufacturers and build and fly these before attempting

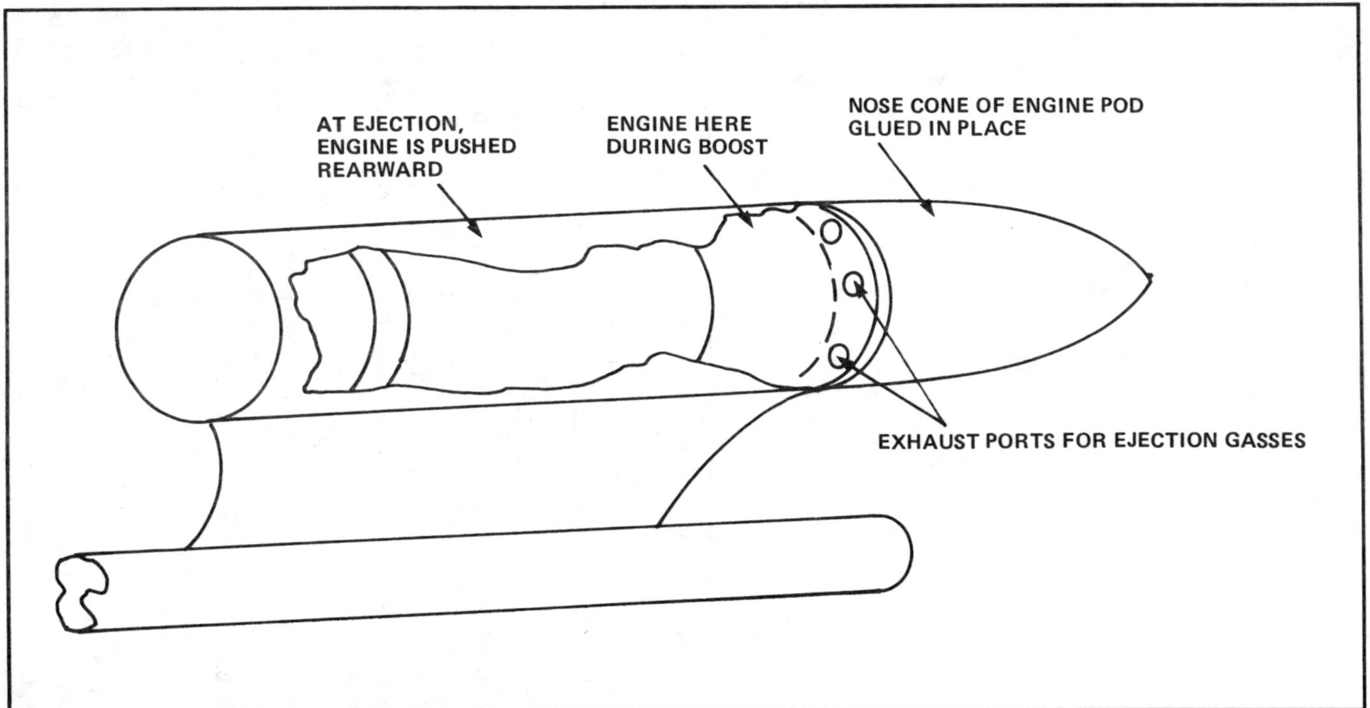

Fig. 5-5. Cutaway drawing of an R/G engine mounting tube.

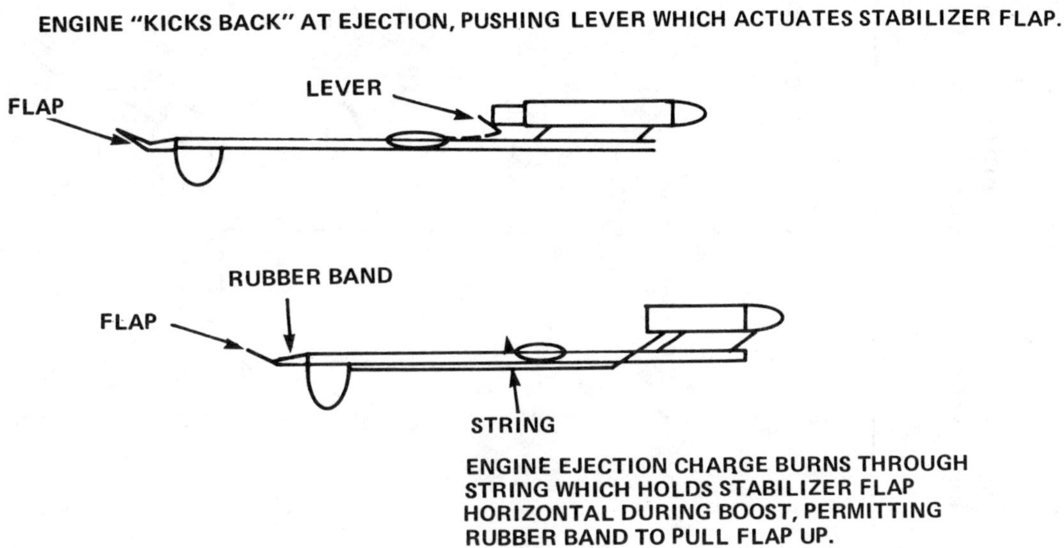

Fig. 5-6. Altering trim of glider at ejection.

your own designs. Even with the kits, you will find that *trimming* or balancing the model for stable flight in both the rocket and glider configurations is a very tricky business.

When flying an R/G, make sure that you have plenty of room for recovery. Don't forget that an R/G is a glider, and it will tend to soar quite a bit before it finally touches down. Of course, you should maximize the recovery potential of an R/G by warping the rudder slightly to the right or the left (as viewed from the top) so that it will tend to travel in a circle. Don't warp it too much, though; too much warp can cause the model to be unstable during its boost phase.

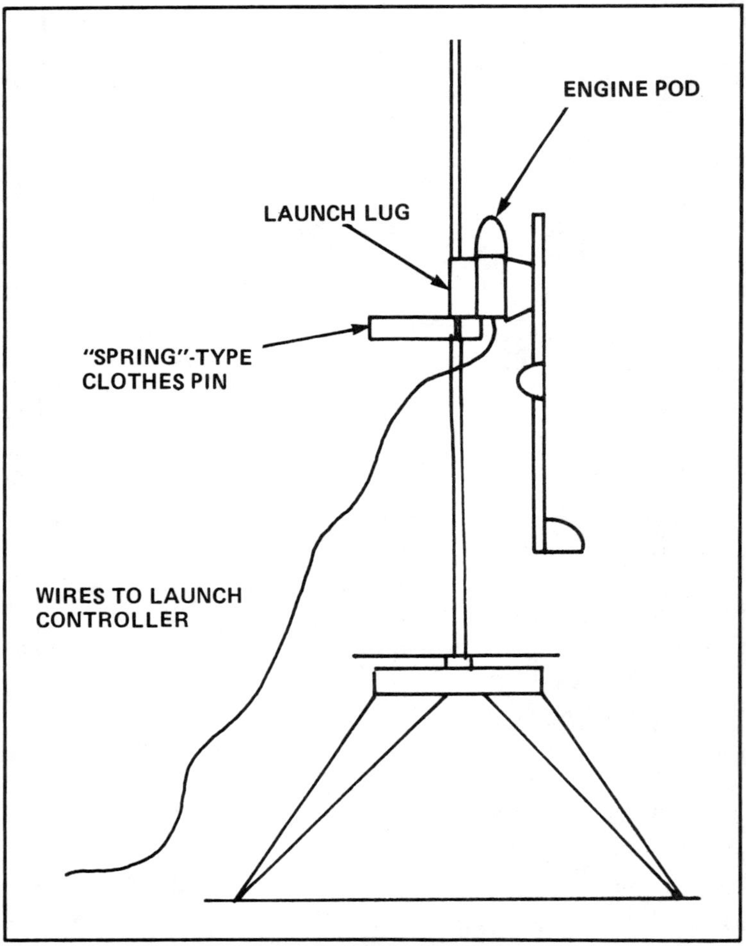

Fig. 5-7. R/G on pad, with clothespin stand off.

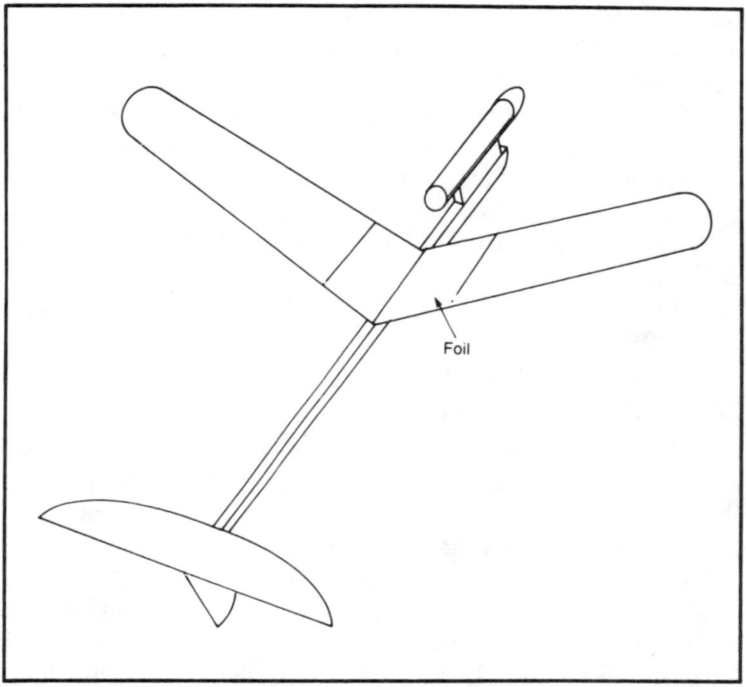

Fig. 5-8. Protect the R/G wing from engine exhaust with foil.

When placing an R/G on the launch pad, use a configuration similar to that in Fig. 5-7. The offsetting of the base of the R/G from the blast deflector will prevent its becoming tangled with the launch controller wires or being knocked off course at liftoff by the launch rod.

It is also advisable to coat the fuselage and wing area closest to the base of the engine with flameproof paint or metal foil to protect it against the engine exhaust (Fig. 5-8).

The visibility of an R/G during flight can be greatly enhanced if the bottom is painted a dark color.

BOOST-GLIDERS

As stated previously, the major difference betweem an R/G and a boost-glider (or, B/G) is the fact that an R/G cannot eject its engine, while a B/G can. The fact that a B/G can eject its engine makes stability less of a problem. Instead of having to devise systems to move the engine toward the rear and/or move control surfaces, you need only take into consideration the amount of weight necessary (in terms of the length of the engine mount boom) to move the CG of a stable glider far enough forward to make it stable during boost. In other words, stabilize the glider with the engine mount boom detached, and then lengthen the engine mounting tube so that the model is stable during the boost phase.

Because you must eject the engine with the B/G configuration, you have to devise a recovery device for the engine. Both of these requirements can be met by using a system similar to that illustrated in Fig. 5-9.

Other than these considerations, the same basic design and flight rules apply for B/Gs as for R/Gs.

R/G AND B/G STABILITY

Now that you have a basic idea of what the basic glider types are, a few words about stability are in order. Figure 5-10 illustrates the basic relationships necessary for stability in both R/Gs and B/Gs. As you can see, the CG at boost (with the engine forward in the case of an R/G, and the boom mounted in the base of a B/G) should be at the leading edge of the wing. The CG during glide should be just ahead of the trailing edge of the wing. This can be tested by balancing the model on the edge of a ruler at the points indicated. If the balance is not true, weight can be added to the nose or tail of the glider, and weight can be added to the boom. Clay works best. You can also change the CG of the boost configuration by lengthening or shortening the engine mounting tube and/or boom.

The glider should, of course, be tested by tossing it. If it is a boost-glider, you should test-glide it without the engine mounting tube and boom. If it is a rocket-glider, you should test-glide it with an empty engine casing in the engine mounting tube, placed in the position in which it will be after ejection.

When you toss the glider, observe its flight carefully. If it dives nose first to the ground, you need to add weight to the tail or remove weight from the nose. If it *stalls* (falls to the ground tail-first or in a horizontal position), you need to add weight to the nose or remove weight from the tail.

Again, all safety rules apply when flying B/Gs and R/Gs.

OTHER GLIDER TYPES

In addition to R/Gs and B/Gs, you will encounter what are known as *parasite gliders* among kits offered by the manufacturers

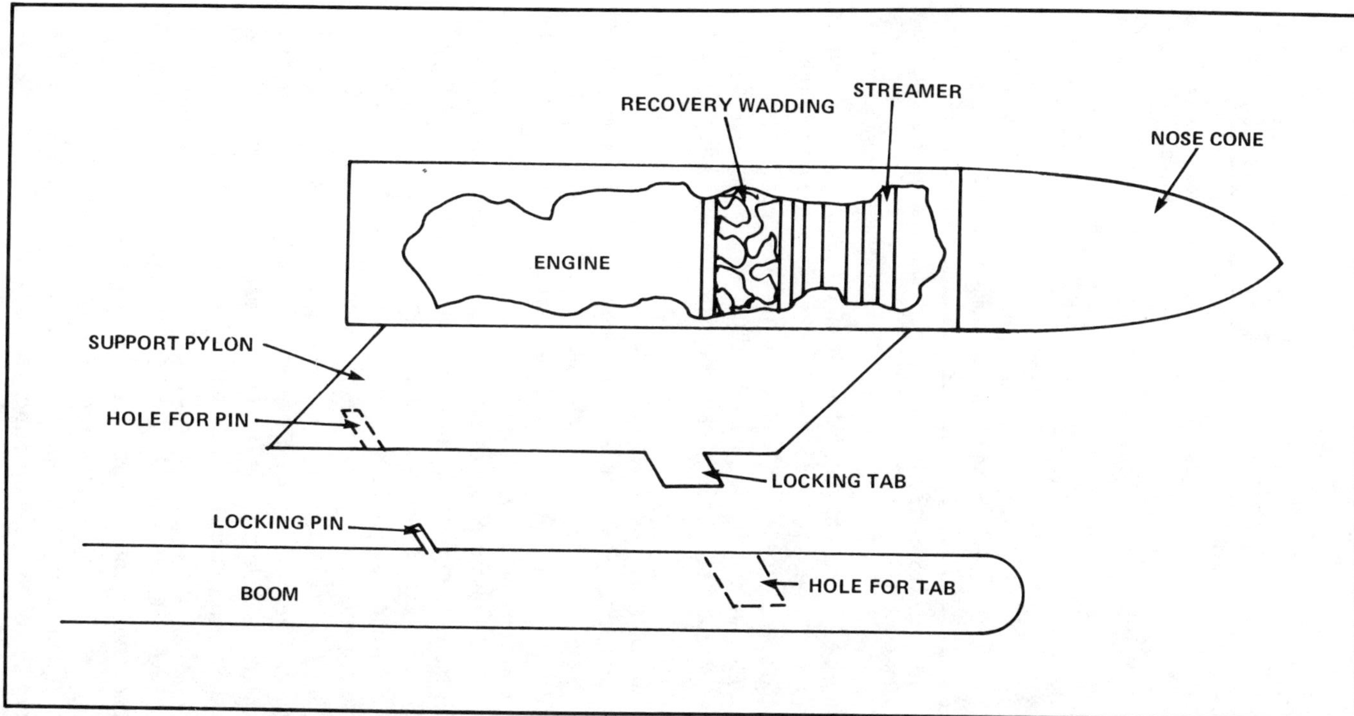

Fig. 5-9. Boost-glider pod with recovery system.

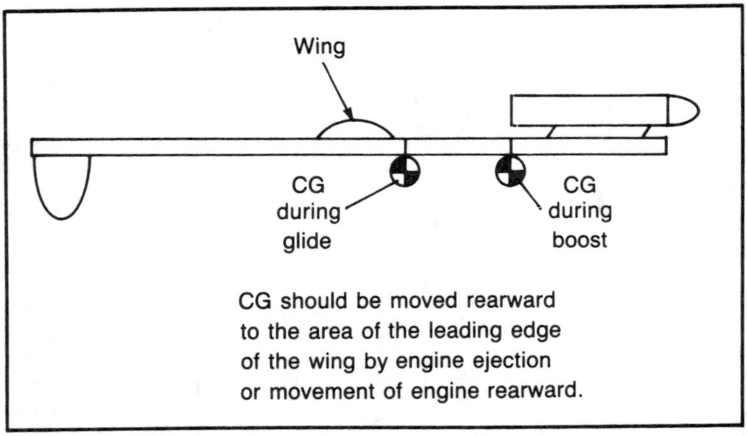

Fig. 5-10. Position of center of gravity of a boost- or rocket- glider during boost and after ejection.

(Fig. 5-11). A parasite glider is just what the name implies—a glider riding as a "parasite" on a model rocket. Parasite gliders may be simple small gliders that hook on to the outside of a large rocket through a launch-lug and dowel arrangement, as in Fig. 5-11, or more complicated model rockets, such as the Estes space shuttle (Fig. 5-12).

Another type of glider is the *flex-wing* (Fig. 5-13). This type rides folded up inside the large-diameter fuselage of a host rocket and is ejected ahead of the parachute or streamer at the rocket's apogee (Fig. 5-14). At this time, the wings of the flex- wing glider are deployed, and it returns to earth separately.

For the purposes of competition and technical considerations, both parasite gliders and flex-wing gliders are considered to be B/Gs. (See Figs. 5-15, 5-16, and 5-17.)

Another type of glider is the result of "crossbreeding" model rocketry and model aircraft techniques. Radio-controlled boost-gliders (RC-B/Gs) have become quite popular in recent years, but the technique of flying them is far from being fully developed. Many rocketeers, including NAR members participating in sanctioned competitions, have attempted both radio control during boost and glide. Unfortunately, the results of this type of flying have been only marginal. Quite often, a good deal of expensive radio and model rocket equipment is destroyed during the first flight of an RC-B/G. There are no second mistakes with this type of model!

Again, flying any type of model rocket using glide recovery must be conducted with utmost safety and caution.

MULTISTAGE ROCKETS

There are several performance-related advantages to flying two- or three-stage model rockets. The most obvious advantage is the fact that it is possible to attain much higher altitudes with multistage rockets than with single-stage rockets (Fig. 5-18). With a two-stage rocket, for example, the model will have already at-

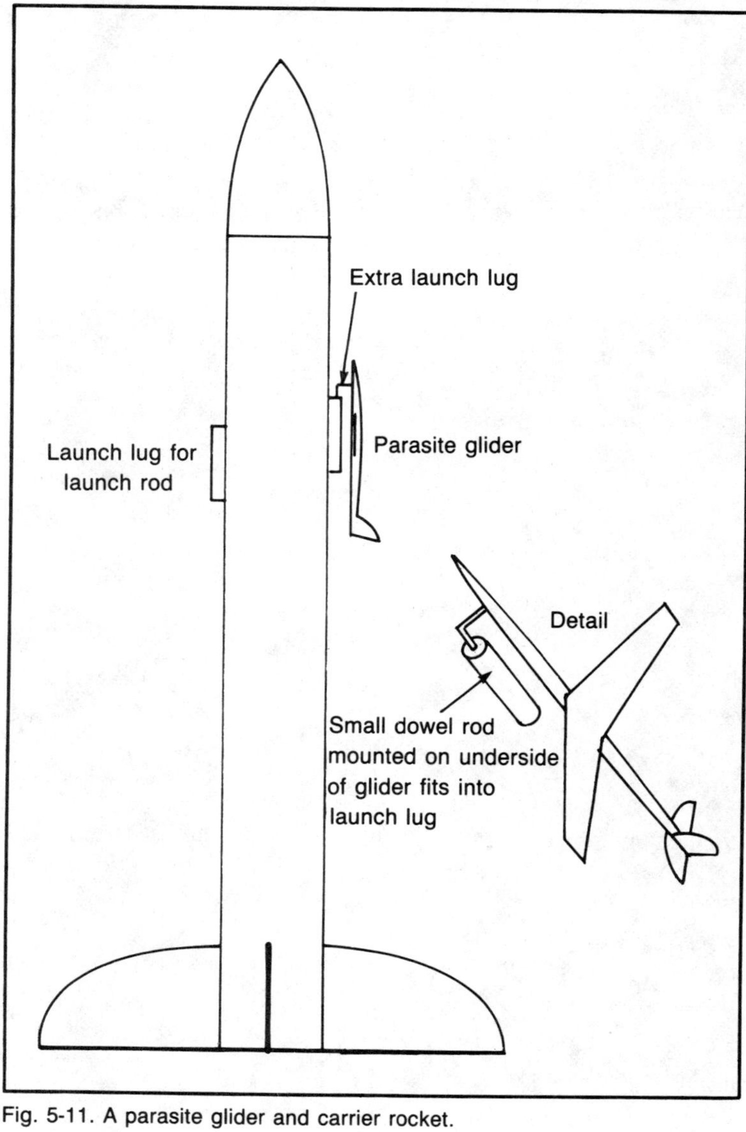

Fig. 5-11. A parasite glider and carrier rocket.

Fig. 5-12. The space shuttles are parasite gliders. Courtesy Estes Industries.

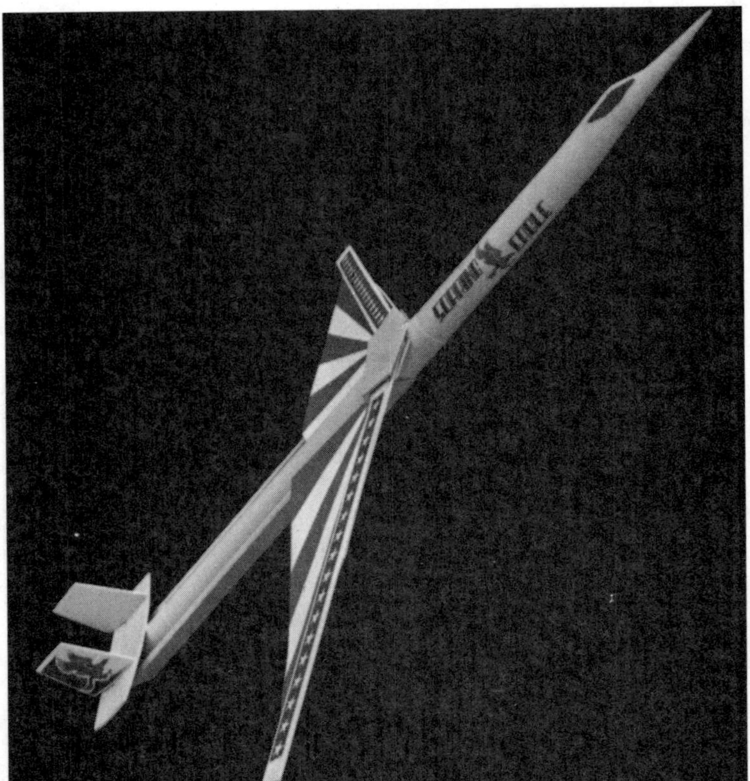

Fig. 5-13. A flex-wing boost glider. Courtesy Estes Industries.

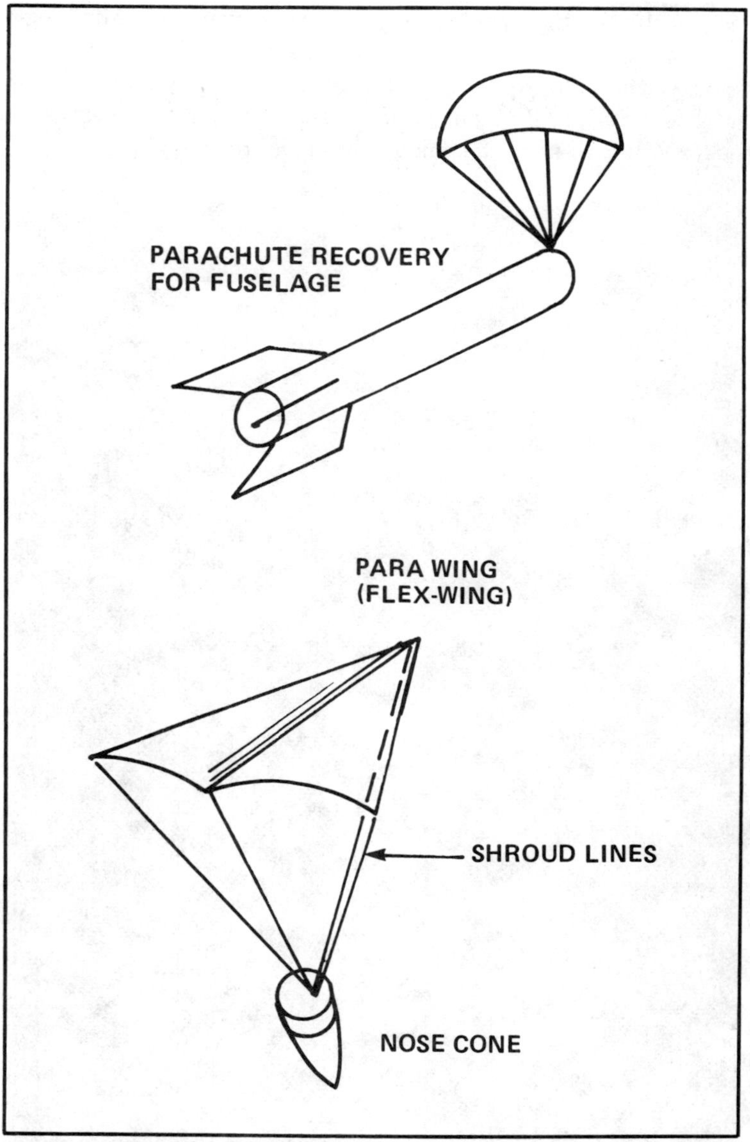

Fig. 5-14. Flex-wing recovery.

tained an altitude of perhaps several hundred feet before the second stage is ignited. At this point, the first stage is discarded, and the second stage continues upward just like any single-stage rocket, with the exception that it is starting its flight high above the ground.

Higher velocities may also be attained with multistage model rockets. Just as the two-stage rocket takes advantage of its "leg-up" in terms of altitude, so does it add to its already considerable velocity when the first stage takes over.

The technique of staging is, of course, derived from professional rocketry, as are many techniques in model rocketry.

Fig. 5-15. Flex-wing B/G on launch pad. Courtesy Tom Kolis.

Fig. 5-16. Rocket-glider on pad. Courtesy Tom Kolis.

In practice, staging model rockets consists of stacking engines on top of one another. When the first engine burns through, its ejection charge ignites the second, as illustrated in Fig. 5-19. Naturally, each engine is enclosed in its own fuselage section, with fins, as with the model pictured here. *Never* attempt to create a multistage model rocket by simply butting engines together.

MULTISTAGE STABILITY

As you may have guessed, adding stages to a model rocket has a considerable effect on its stability. Because additional engines mean extra weight toward the rear of the rocket, an extra stage

Fig. 5-17. Boost-glider on pad. Courtesy Tom Kolis.

Fig. 5-18. A multistage rocket. Courtesy Estes Industries.

or two moves the CG rearward, which is an undesirable effect. To compensate for this effect, the lower stages on a model rocket must have progressively larger fin areas, so that the CP is also moved to the rear.

As the stage or stages drop off, the weight is removed, and the CG is moved forward. The fins go with the stage sections, so the CP is likewise moved, adjusting the stability of the rocket in flight.

The larger rear fin area can be a source of problems despite its benefit in increasing stability. The problem with large (very

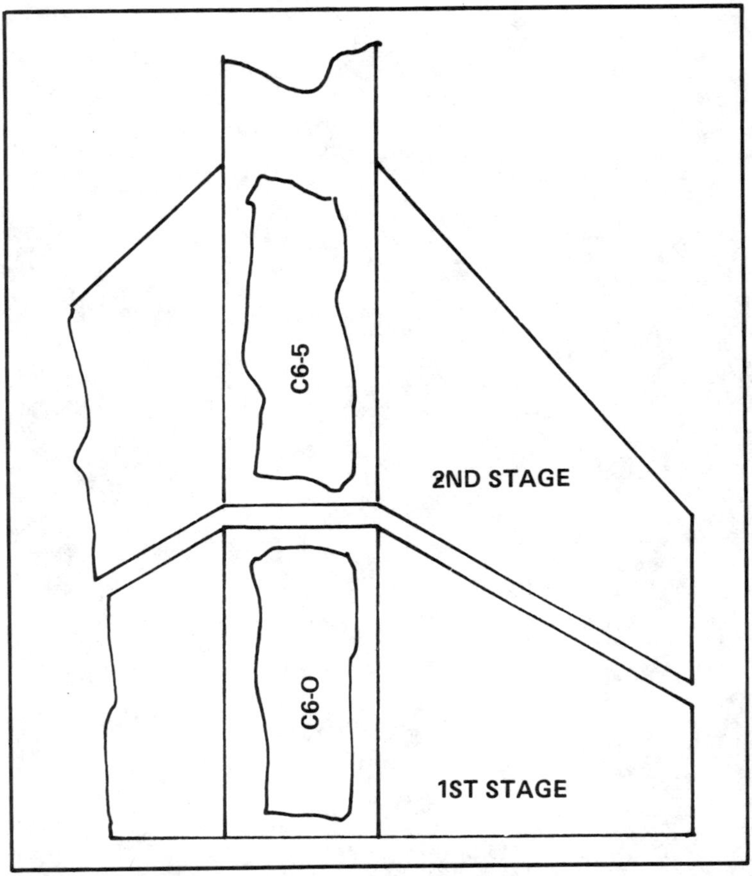

Fig. 5-19. Engine placement in a multistage model.

large) rear fins is the fact that they make the rocket more susceptible to weathercocking (tilting into the wind). Even the slightest breeze can cause a rocket to weathercock, and the effect is amplified greatly in multistage rockets. Therefore, it is inadvisable to fly multistage rockets on windy days. You should also take care not to tilt the launch rod too far from vertical, because the tilt angle will be multiplied as the rocket boosts. A three-stage rocket can easily assume a horizontal flight by the time the third stage ignites, which can result in its quickly disappearing over the horizon (or worse), so take care.

CLUSTERING ENGINES

Clustering model rocket engines (Fig. 5-20) involves combining two, three, or four engines together in one stage and igniting them simultaneously. In addition to providing a lot of smoke and noise, clustering engines will greatly increase the lifting capacity of a model rocket. Some heavier model rockets, such as the Saturn V models offered by Estes and Centuri, need the extra boost of clustered engines to achieve a satisfactory flight. Other rockets, especially those used in payload applications in research or competition, need the boost to get that weight off the pad.

The most important aspect of engine clustering is ignition. It is imperative that all engines in a cluster ignite at the same time. If they don't, the stability of the rocket could be ruined. This is because the engines in a cluster are not individually centered in the fuselage, and off-center thrust will totally throw off stabilizing influences such as fins.

Figure 5-21 illustrates several methods of cluster ignition, all

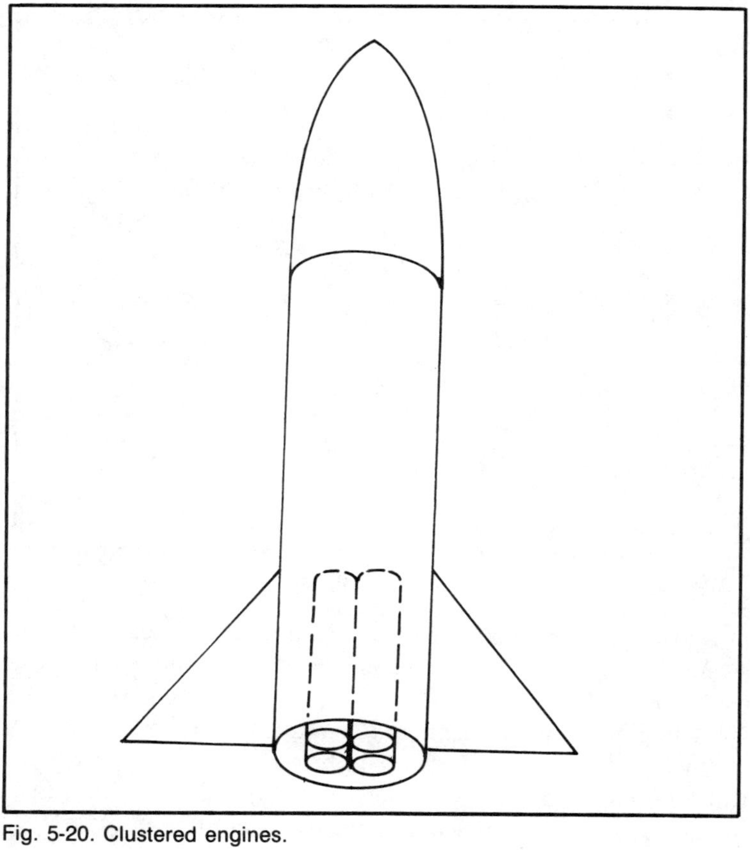

Fig. 5-20. Clustered engines.

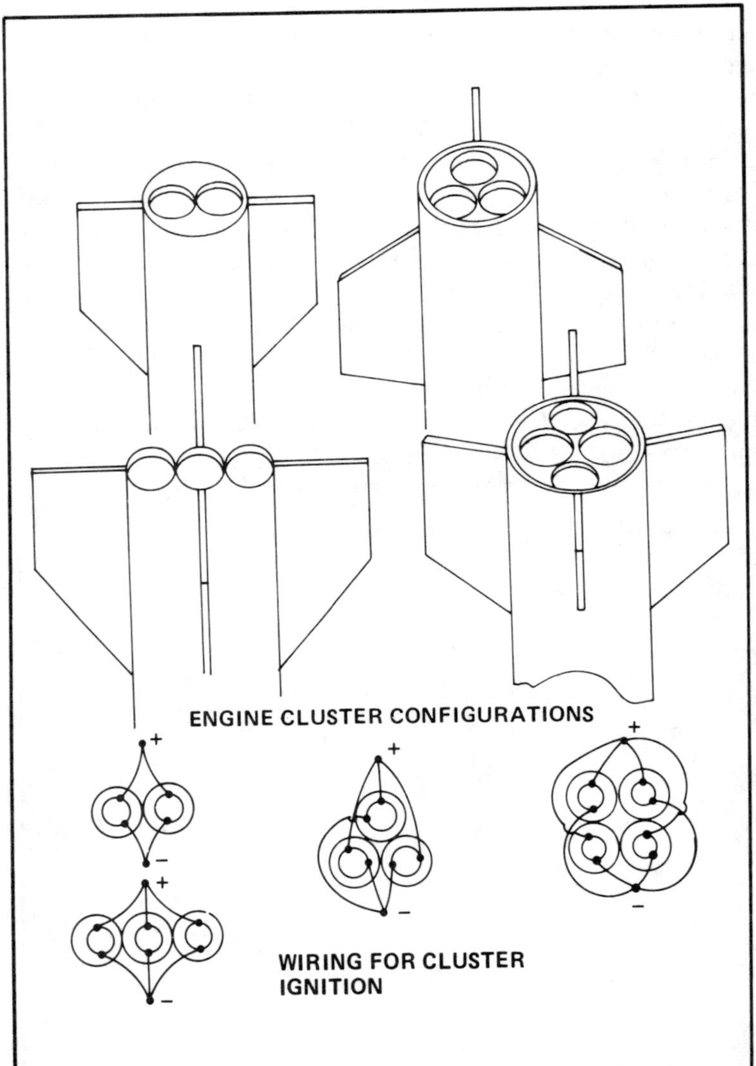

Fig. 5-21. Cluster configurations and wiring.

proven by years of use by experienced model rocketeers. Do not attempt to use fuses to ignite engine clusters (or even single engines)!

When using several igniters to ignite clusters, be sure to use a 12-volt battery instead of a 6-volt battery, because the current will have more resistance to overcome (Fig. 5-22). Be sure, too, to only use the same types of engines in a cluster.

HIGH-POWER ROCKETS

The generally accepted definition of a high-power model rocket is any rocket that has over 20 newton-seconds of total impulse. In other words, that's any rocket using more than one D engine, or an E or F engine, or any combination of engines with a total impulse over 20 newton-seconds (Fig. 5-23).

The first word in flying high-power rockets is *safety*. In addition to being more powerful than the average model rocket, a high-power rocket is generally heavier and sturdier than a kit off the shelf. Because of this, high- power model rockets should be handled with more than a little respect.

The high average impulse levels of high-power rockets dictate a strong construction. Therefore, you should use epoxy glues, as

Fig. 5-22. Cluster rocket lifting off. Courtesy Tom Kolis.

Fig. 5-23. High powered rocket in flight. Courtesy Tom Kolis.

opposed to the white glue you may be using to build smaller rockets. You should employ such strengthening techniques as running the fins right through the fuselage and anchoring them to the engine mounting tube (Fig. 5-24). If you use this technique (along with liberal quantities of epoxy for bonding and filleting), you won't have to worry about fins shredding or breaking off at liftoff (it has happened!).

Careful selection of materials is also important. Balsa is not strong enough for high-power applications, but birch plywood (the type sold for model airplane construction) and basswood are. Body tubes with walls thicker than those offered by the large manufacturers are also useful; these can be obtained from some smaller manufacturers who specialize in high-power applications.

When constructing a high-power model rocket, it is a good idea to coat the fins and nose cone with epoxy. This will not only strengthen them for flight, but more importantly, provide added

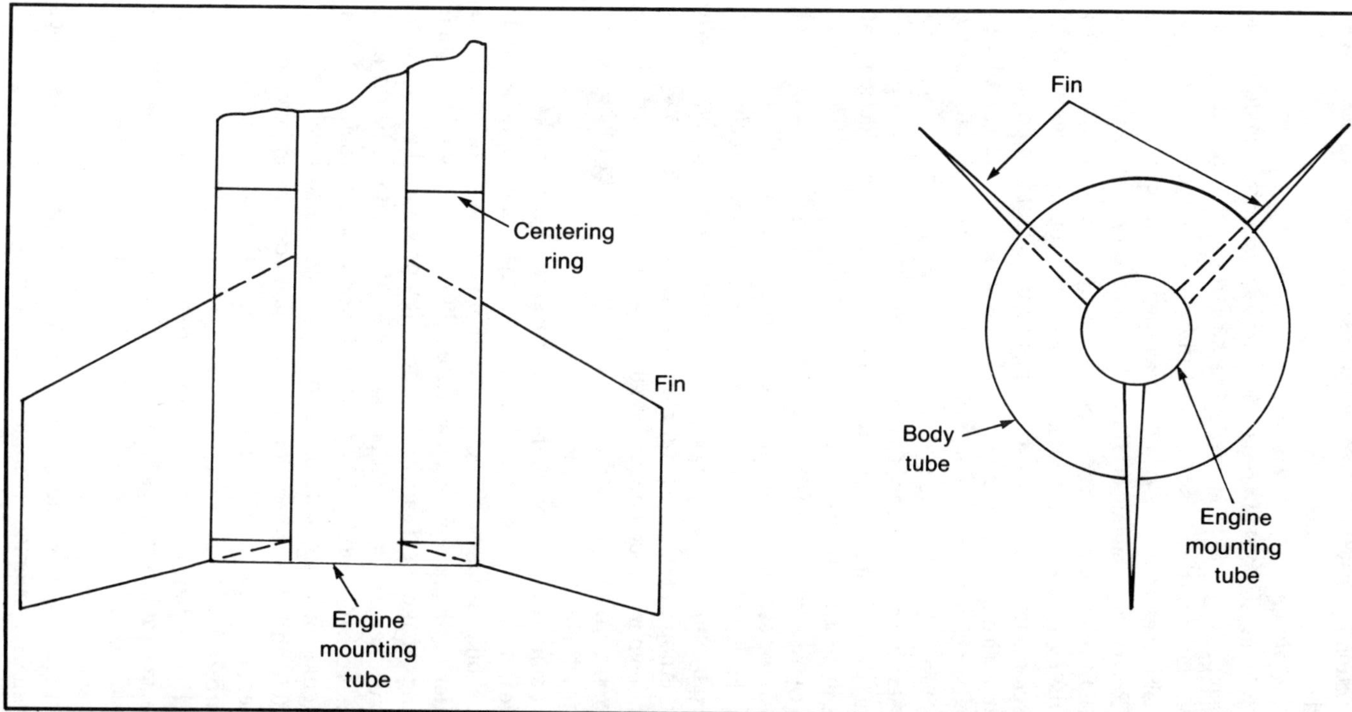

Fig. 5-24. Strengthening fins by running them through body tube.

protection against the increased forces of recovery impact caused by the rocket's greater weight.

Flying high-power rockets requires a bit more planning than the average amount of preparation for flying average models. Obviously, a larger field is needed because the higher the apogee of a rocket, the farther it is likely to drift (with some high-power rockets, altitudes in excess of 3000 feet can be expected.)

Recovery technique is critical. It is necessary to bring the rocket down from its high altitude as quickly as possible to minimize drift, but the extra weight can cause the rocket to drop too quickly. Most rocketeers use large (at least 6 inches by 10 feet) streamers, usually coated with silver mylar on one side. Not only does such a large streamer slow the rocket enough for safe recovery without permitting it to drift, but it also makes the rocket easy to spot. Along with the streamer, rocketeers also use a quantity of powdered tempra paint, packed into the nose, to create a large cloud on ejection that aids in spotting the rocket.

Extra large designs that are somewhat fragile and thus need a slower descent can be safely recovered on calm days with large parachutes (3 feet or more in diameter). The plastic parachutes provided by most model rocket manufacturers are not strong enough for this application, however. Parachutes made using rip-stop nylon with sewn-in shroud lines are ideal. If you are unable to make these for yourself, the manufacturers offering high-power rocketry can supply them.

Launch systems for high-power rockets, like the rockets themselves, must be extra strong. Instead of the standard launch rod, you should use a rod 6 feet long and 1/4-inch in diameter. Small sections of body tube, such as Estes BT-5, or empty mini-engine casings make ideal launch lugs for high-power rockets (as with every other part of a high-power rocket, these should be fastened to the rocket securely). The blast deflector for a high- power launch system should be made of heavy gauge steel, and ideally should include an L-shaped section of small diameter heating conduit to deflect the exhaust of the engine(s) from vertical to horizontal. (See Fig. 5-25.)

Remember, too, that no model rocket may weigh more than 16 ounces at liftoff. This includes, of course, the high-power variety we are discussing here. This isn't as much of a restriction as you might think, though.

Figure 5-26 is a design for a typical high-power rocket. Note the special techniques employed in its design.

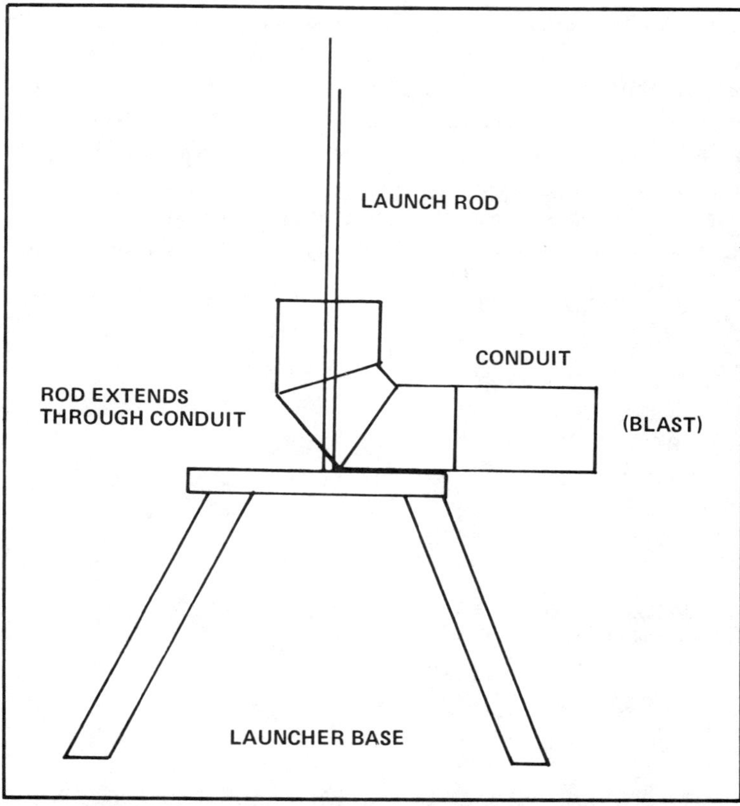

Fig. 5-25. Launch pad and blast deflector for high-power applications.

Use a little common sense and caution when flying high-power rockets, and you'll enjoy many successful flights.

RECOVERY DEVICES

You are already familiar with parachute, streamer, and glide recovery techniques, as they have been discussed earlier in this book. There are a few other recovery techniques, which we will discuss here.

Two alternative recovery techniques are variations on B/Gs. It is possible, for example, to use a flex-wing glider to recover an entire rocket by keeping the rocket attached to the flex-wing with a length of shroud line (affixed to the flex wing at its center of gravity or point of balance). This emulates the *rogallo wing recovery* method that NASA engineers considered as an alternate to parachute recovery of early space capsules. The wing, in effect, changes shape during flight. With this technique, the rocket will be carried to the ground in a spiral pattern.

Another variation on B/G is the *swing wing* (Fig. 5-27) concept, with which the manufacturers experiment from time to time. A swing wing carries a wing folded along its length, and the wing is snapped outward by the ejection charge (the ejection charge releases a string holding back a rubber band, which actually moves the wing). There is a great amount of difficulty in making one of

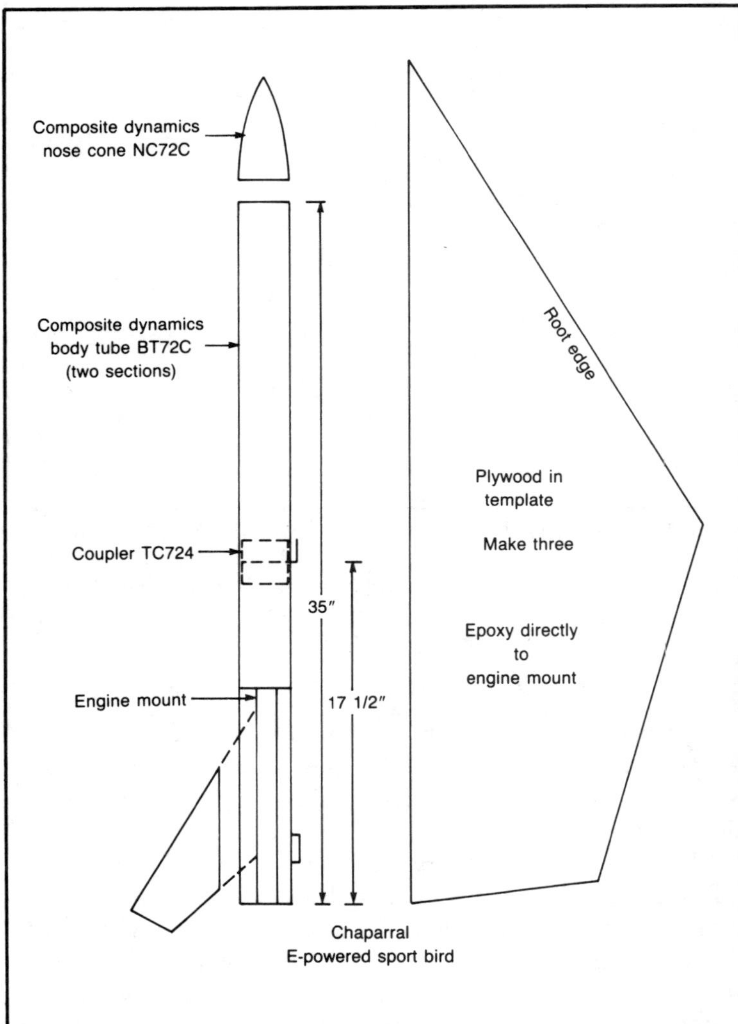

Fig. 5-26. The Chaparral, an E-powered rocket designed by Rich Cardillo.

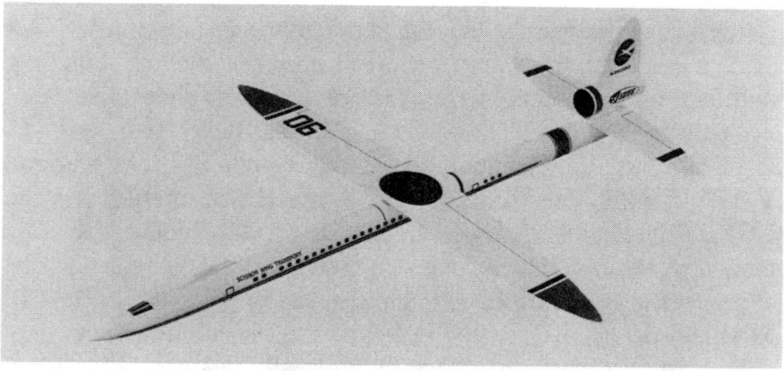

Fig. 5-27. Swing wing boost-glider. Courtesy Estes Industries.

these recovery systems work, however. Even the kits using swing wing recovery are notorious for their poor performance.

A very popular alternate method of recovery is the *gyro*, or *helicopter recovery* technique. This is approached in more than one way. The standard technique is to attach three folding, spring-loaded rotors to the base of the nose cone. The rotors are folded back and inserted in the rocket fuselage when the rocket is prepped for launch. When the engine ejection charge is ignited at apogee, the nose cone and rotors are ejected. The springs attached to the rotors pop the rotors out, and they begin to spin. The rotors should have airfoil shapes, like airplane propeller blades, so that they will turn with air movement. The resistance of the air flow as the rocket drops toward the ground will cause the rotors to spin, and this action creates drag that slows the rocket's descent.

An alternative to putting the rotors inside the fuselage is to construct the rotors so that they fold down to form the fuselage of the rocket. This is an extremely difficult technique, but is successfully employed in rockets built by the Roto-Rock Company.

Another helicopter recovery technique changes the surface of the fins so that they make the rocket spin. The change occurs after engine ejection and is activated by the engine leaving the rear of the rocket. The change is minor—two small flaps, one on each fin are flipped up in opposing directions. Estes Industries' "Gyroc" kit is the best performer in this area.

SCALE MODELING

Scale modeling (Fig. 5-28) is the most rewarding model rocketry activity for the craftsman. The aim of scale modeling is to produce an exact duplicate of a real space or aerospace vehicle, with performance parameters as close to the prototype as possible. NAR scale modelers often spend a year or more researching, building, and testing their scale models intended for competition.

Obviously, only rockets or missiles that have an appearance of stability are good candidates for scale modeling. The German V-2 (Fig. 5-29), the United States Army Hawk missile, and the USAF Phoenix air-to-air missile are all good candidates for scale modeling, because they have in common a fairly large rear fin area, along with a fuselage long enough to keep the CG ahead of the CP. Missiles such as the Titan, Snark, and the Poseidon are poor candidates for scale because they are inherently unstable (Fig. 5-30). They are stabilized in flight by techniques such as spinning, or use inertial guidance systems that change the angle of thrust or move control surfaces, all unavailable for use by the model rocketeer.

With this in mind, it is a good idea to build a few scale kits before trying to design your own. The various manufacturers offer a wide range of scale kits, ranging from the Canadian Black Brant and the United States space shuttle to the Mercury Redstone and the Saturn V (Fig. 5-31). Building kits such as these will give you a feel for the craftsmanship required for scale modeling (Fig. 5-32).

Once you've gained experience building some of the more difficult scale kits, you may feel confident enough to design and build your own scale model. The first thing to do is to select a rocket

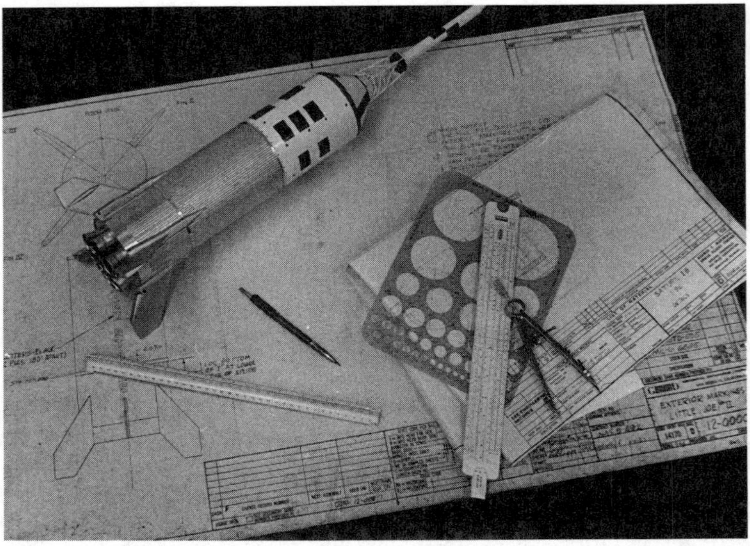

Fig. 5-28. An exquisite example of scale modeling. Courtesy Centuri Engineering.

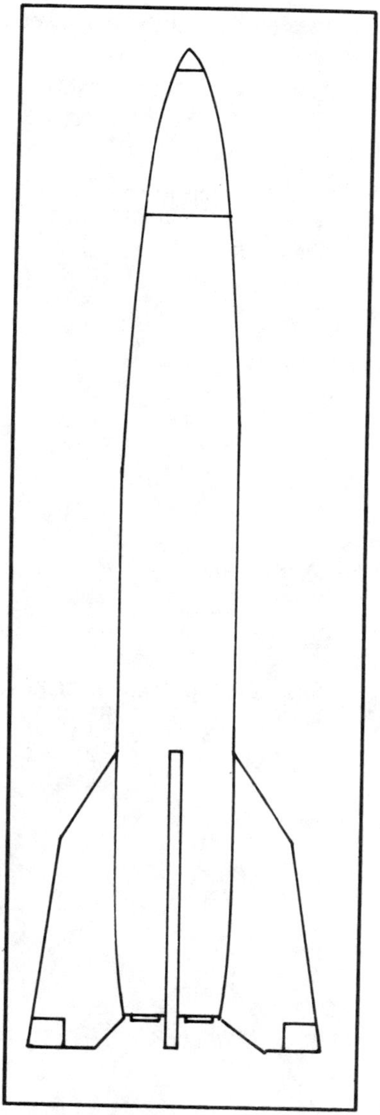

Fig. 5-29. The German V-2 rocket's clean lines make it easy to model.

or missile to model, of course. Again, try to choose one that is obviously stable as a model rocket, which means it should have a large rear fin area and a CG ahead of the CP.

Once you have selected the rocket you are to build, you should begin researching it. Go to the library and look for references to your "bird" in encyclopedias and books on rockets and missiles. Make photographs of any plans, drawings, or photos of the rocket you may find, and note the color schemes and insignia the rocket carries.

The next step is to try to obtain color photos of the rocket. If you happen to live near an air force base or a military or aviation and space museum (such as the United States Air Force Museum or the Smithsonian's National Air and Space Museum), you may well find the rocket or missile you wish to model on display. If you don't have access to such a resource, the next best thing is to write to such museums and ask if they have photos and/or books about the rocket or missile in which you are interested for sale.

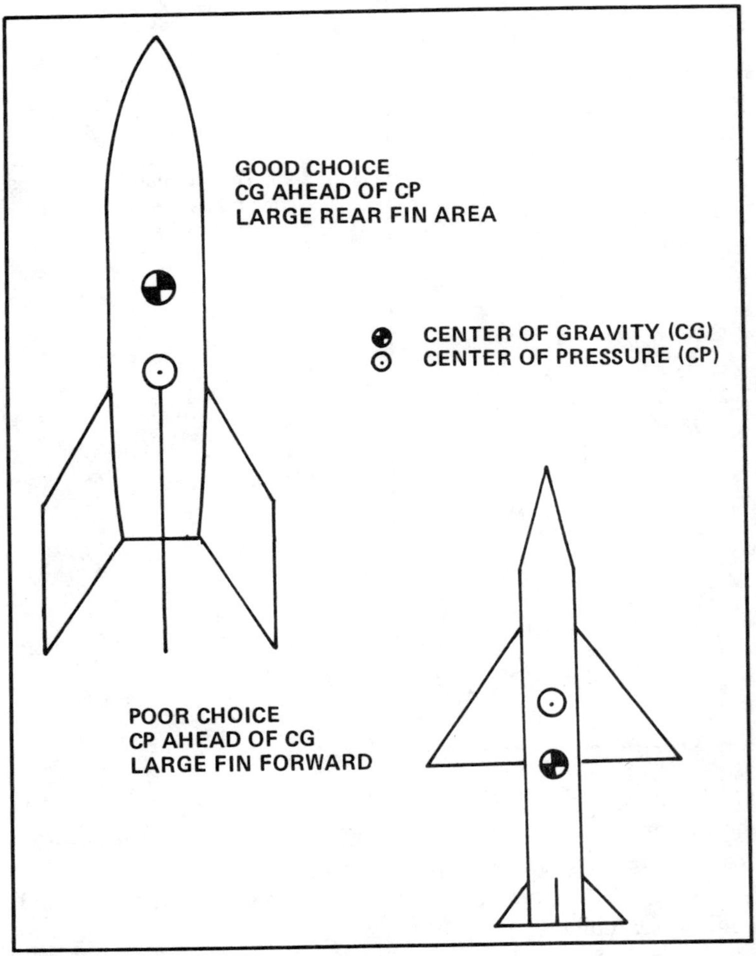

Fig. 5-30. Choosing a rocket to model.

Fig. 5-31. A scale kit of the Saturn V. Courtesy Estes Industries.

Fig. 5-32. The ultimate in scale, a rocket and launcher. Courtesy Tom Kolis.

A great deal of information may also be obtained from the manufacturers of rockets or missiles. Companies such as Raytheon, Thiokol, and General Dynamics will usually be happy to provide you with necessary information such as dimensions, color schemes, and accessory equipment, as well as photographs, as long as the rocket or missile in question is not classified. Books or articles on rockets and missiles of a specific type almost always list the manufacturer of the rocket.

Once you have assembled the necessary photos, dimensional and color data, and any other information you feel necessary, you will have to decide on the size of your model. If you have the necessary measuring equipment, you can build the model on a specific ratio of model size to size of the prototype. For example, if the prototype is 40 feet in length, and you wish to build on a scale where 1 inch of length (or any other dimension) of the rocket is equal to 1 foot of length of the prototype, you would be building a model 40 inches in length. You would scale every other feature and dimension of the model accordingly (Fig. 5-33). In some cases, exact scale would require that you measure parts to 1/10th of an inch, or even less, so calipers are a definite plus (Fig. 5-34).

Selecting materials is the next step. Obviously, the diameter of body tube used to construct the fuselage is critical, because the entire model's dimensions will be based on this. Fortunately, there are many different sizes of body tubes offered by the various manufacturers, so you should not have too much trouble finding one with the proper diameter to match the scale you have selected. If you cannot find a body tube with exactly the same diameter as your scale requires, it is possible to "build up" the outer diameter of a body tube by wrapping it with paper or *silkspan*. If all else fails, you can make your own body tube, as described in Chapter 2.

Fig. 5-33. The Estes Cruise Missile is a difficult scale project. Courtesy Estes Industries.

Fig. 5-34. The space shuttle is a popular scale project. Courtesy Estes Industries.

Building a scale model should be carried out in two stages. The first stage involves building a "boilerplate" model—one that is a rough approximation of the finished design, and which can be flown to test stability and performance. The boilerplate model need not be painted or finished to the specifications of your design, as long as it is the same size and shape as the final version. Testing your design with a boilerplate model can save you time and effort that might be wasted on building a completely finished model that proves unstable.

Construction techniques, materials, and tools used in scale modeling can be the same as those used in building any model rocket. It is advisable, though, that you obtain the best possible saw, razor knife, and paintbrushes available for finishing. An airbrush is preferable to spray painting for the overcoat, and you will want to use decals for lettering and numbering whenever possible, to add to the realism of your scale model.

Try to duplicate the performance of the prototype with your scale model, but don't overdo it. You can, for example, build a scale model of a Saturn V from a kit or from scratch, but you will find it impossible to stage the model, because the Saturn V has no fins above the first stage. The upper stages were designed to perform where there is little or no air and are stabilized with inertial guidance systems.

Once you've successfully built and flown a scale model, from a kit or from your own designs, you'll find that you have something of which to be quite proud (Fig. 5-35).

PLASTIC CONVERSION

Converting plastic models to fly is a very challenging activity. Not only must the problems of stability and safe recovery be overcome (because plastic models are more fixed in shape and heavier than conventional model rockets), but also the problem of how to mount the engine, recovery system, and other necessities must be dealt with.

The simplest subjects for plastic conversion are those rockets that are already stable, by virtue of having a large rear fin area and a long fuselage (this translates into a CP to the rear and a CG to the front.) Ideal in this context is Revell's V-2, first issued in the 1960s, and later re-issued (in 1982) as one of a series of "History Makers."

The V-2, as shown in Fig. 5-29 has a large rear fin area and a reasonably long fuselage. It is, in fact, the "classic" rocket design and is very close in profile to several of the model rocket manufacturers' beginners' kits. Thus, it is already stable, in theory.

In practice, a converted plastic V-2 will probably need some nose weight—a small amount of clay put into the nose cone—to move the CG forward. The need for this can be determined by swing-testing, as described in Chapter 2.

Fig. 5-35. Enhancing a model with a background. Courtesy Estes Industries.

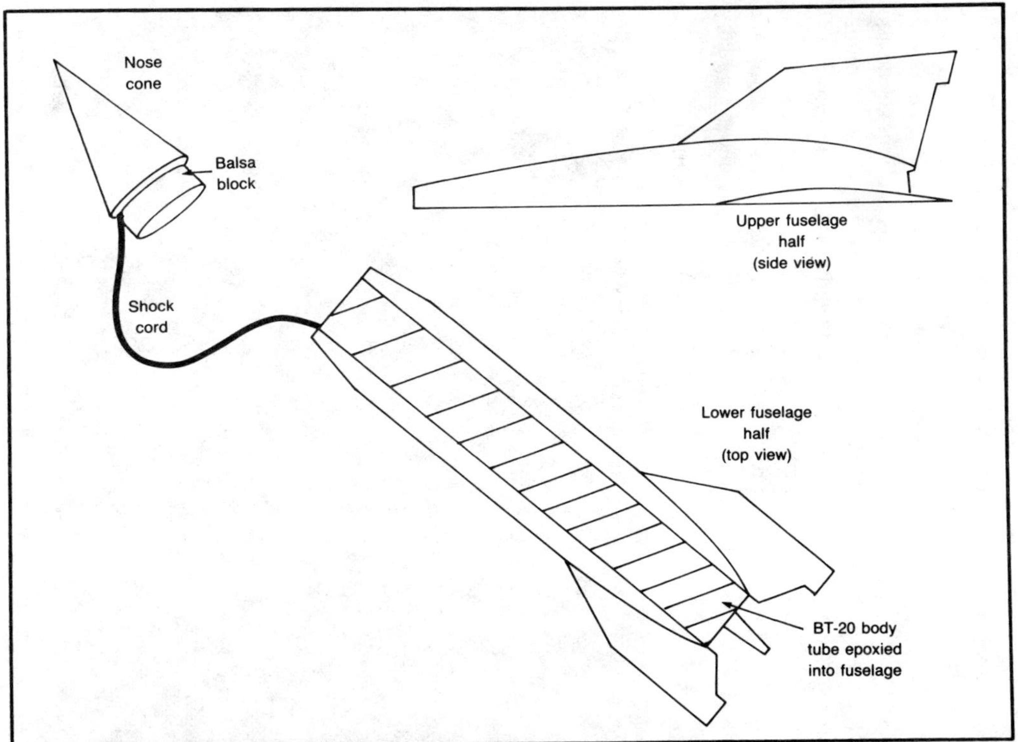

Fig. 5-36. Plastic V-2 Conversion.

Because the Revell V-2 is molded in two sections, as shown in Fig. 5-36, it is relatively simple to locate the engine and recovery system. Actually, the conversion of the plastic V-2 may be viewed as putting the V-2 around a model rocket.

To convert the V-2, follow the diagrams in Fig. 5-37, first separating the nose cone and attaching the shock cord to it. Next, cut out the plastic spars joining the two fuselage halves to permit the insertion of an Estes BT-20 (or similar) body tube. The body tube should extend from the base of the rocket (Fig. 5- 38) to the top of the fuselage halves and should be glued into place with epoxy on one of the halves. The remaining fuselage half should then be glued to the body tube and the other fuselage half and set aside to dry.

You may wish to fix a small balsa engine block, or *bulkhead*, to the base of the nose cone so that the nose cone will stay on the rocket during flight. The piece of balsa should extend 1/4-inch into the body tube, but should not be cemented into place until you have determined whether or not you will need to add clay nose weight to the nose cone.

The shock cord should be attached to the inside of the body tube with a shock cord anchor, and a parachute or large streamer should be attached to the nose cone.

A *thrust ring* should be placed inside the rear of the body tube. Place a small amount of glue entirely around the inside of the body tube about 2 inches from the base, and use an engine casing to push

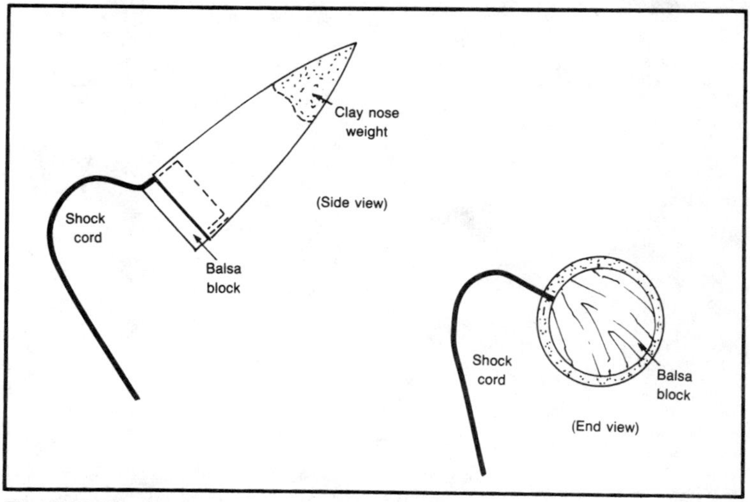

Fig. 5-37. Nose cone detail of the V-2 conversion.

the thrust ring into the body tube until the base of the engine casing is flush with the end of the body tube (Fig. 5-39).

Let the entire assembly dry and then finish and paint the V-2 as you wish. When swing testing the model for stability, be sure to have an engine in the base, because you must simulate the actual flight conditions of the model for a swing-test.

When flying this model, wrap a couple of turns of masking tape around the base of the engine, so that the engine will stay in the rocket when the ejection charge is ignited.

Other models, such as the Viper fighter from "Battlestar Galactica," the Saturn V, and the F-104 Starfighter aircraft, have been successfully converted to plastic flying models, but they are very difficult projects. The Viper is converted using techniques similar to those used to convert the V-2, while the Starfighter uses a *rear-ejection system*, wherein the engine and its mounting assembly

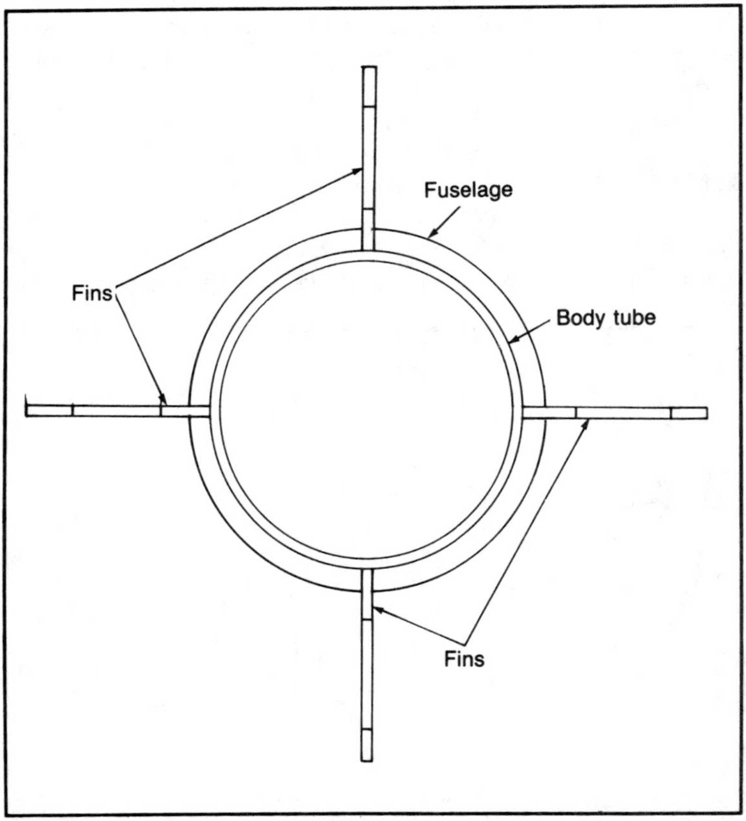

Fig. 5-38. Base of plastic V-2 conversion (end view).

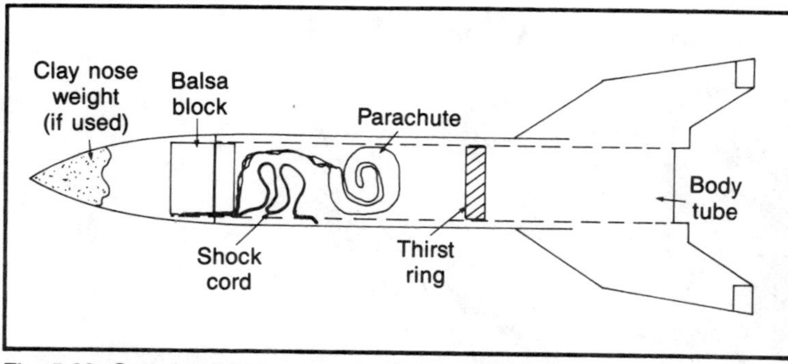

Fig. 5-39. Completed V-2 conversion in cross section.

are ejected rearward at apogee, pulling a large parachute out of the fuselage.

It is suggested that you build and fly many different types of model rockets before attempting even simple plastic conversions. The necessary building and flying expertise must be acquired by experience.

When flying a plastic conversion of any type, be sure to use only engines that have an initial high impulse. For example, with the V-2 conversion, you will want to use a C6 engine, rather than a C5, because the C6 has a higher initial and average thrust, even though it has a shorter burn time. (The burn time of the C6 is approximately 1.66 seconds, compared to 2.5 seconds for the C5. The average thrust of the C6 is 6 newton-seconds, while the C5's average thrust is 5 newton-seconds.)

Also, because a plastic conversion is a heavier-than-average model rocket, you should use an engine with a shorter delay time than you would for a model rocket made with conventional materials (a C6-3, instead of a C6-5, in the case of the V-2).

Again, safety is the watchword here. Double check your design, your construction, and your flying conditions, and test-fly your conversions with as few people present as possible.

Chapter 6

Model Rocket Performance

This chapter will deal with a variety of techniques for measuring and predicting the performance of model rockets, as well as alternative methods of stabilization that can affect performance.

ALTITUDE TRACKING METHODS

The basic requirements for an *altitude tracking system* are a method of measuring the angle of elevation between a model rocket in flight an an observer on the ground and the distance between the observer and the launch pad.

The angle of elevation can be measured with an *altiazimuth scope (theodolite)* (Fig. 6-1), or with an easily constructed device made from a cardboard tube, a protractor, a length of string, and a weight, as illustrated in Fig. 6-2.

The simplest method for determining the altitude of a model rocket using the theodolite or angle measuring device is to find the angle of elevation, as illustrated in Fig. 6-3. Once you have measured the angle of elevation, you can easily find the altitude of the rocket by multiplying the tangent of the angle by the distance of the observer from the rocket. For your convenience, Table 6-1 lists tangents of the angles you will be dealing with in measuring model rocket altitude.

This method isn't always totally accurate because of the fact that the rocket may be weathercocking, and the observer may be measuring a false angle. To overcome this potential problem, you

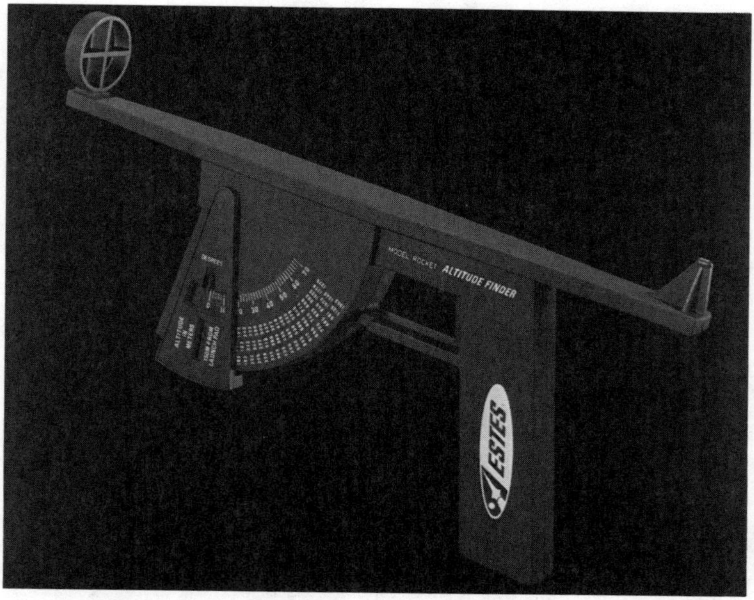

Fig. 6-1. The Estes Altitrak. Courtesy Estes Industries.

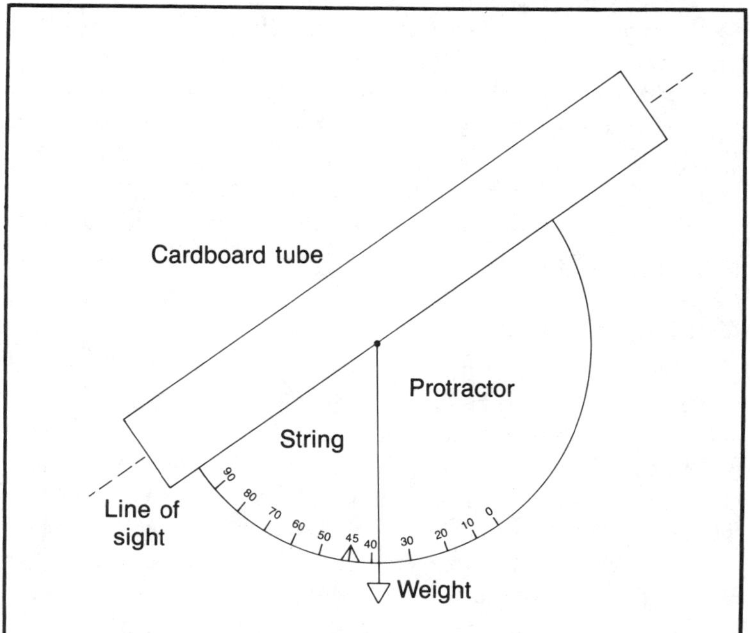

Fig. 6-2. A simple homemade tracking device.

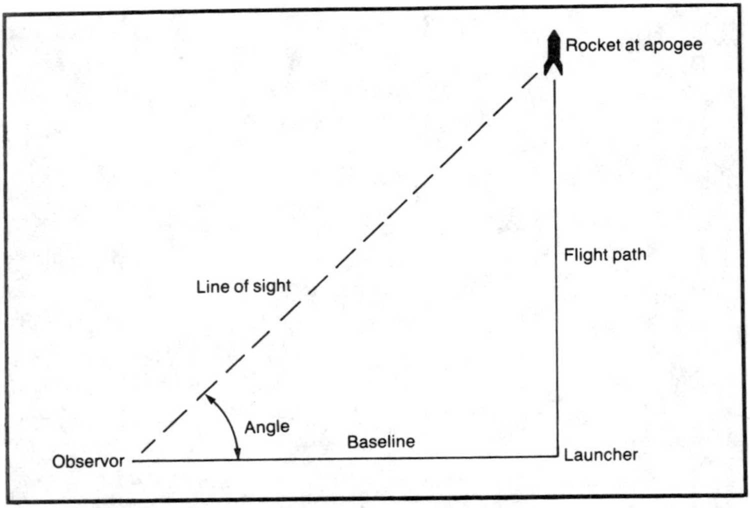

Fig. 6-3. Tracking a rocket uses simple geometry.

should be sure that the line between the observer and the launch pad is perpendicular to the wind, as in Fig. 6-3.

Predicting the altitude that your model rocket will reach at apogee requires several calculations, including center of pressure, which are quite complex. For this reason, although I will discuss some of the necessary formulae here, the interested rocketeer is referred to the following publications: *Altitude Prediction Charts* (TR-10, available from Estes Industries); *Calculating the Center of Pressure* (TIR-33, available from Centuri Engineering Co.); and *Model Rocket Altitude Performance* (TIR-100, available from Centuri Engineering Co.).

AERODYNAMIC DRAG

Aerodynamic drag (the resistance of air as a model rocket passes through it) is one of the most important factors affecting model rocket performance.

As a model rocket moves through the air, it collides with millions of air molecules, and each molecule has some small effect on the rocket. Together, these molecules can cause rather significant changes in the speed and direction of a rocket, so their effects may be taken into consideration in the design of any model rocket.

The effect of air flow on a model rocket can be most easily demonstrated by holding your hand outside a moving car. If you hold your hand with the palm facing the direction of the air flow, your hand will encounter a large amount of resistance and will be pushed back. If you present the edge of your hand to the air flow, there will be less resistance, and your hand will not be pushed with as much force.

The reason for the difference in force is the larger area presented by your palm to the air flow. The air has a surface area on which to act, and the force is correspondingly greater.

In terms of model rocket design, it is always the intent to present the smallest possible area to the air flow around the rocket, and thus create the smallest amount of drag. It is for this reason that model rocket nose cones are rounded, instead of flat.

Figure 6-4 shows various types of model rocket nose cones, with their relative efficiency in reducing aerodynamic drag. The slightly rounded, or *parabolic,* shape is actually the best design. At

Table 6-1. Table of Tangents.

ANGLE	TANGENT	ANGLE	TANGENT	ANGLE	TANGENT
1	0.017	31	0.601	61	1.80
2	0.035	32	0.625	62	1.88
3	0.052	33	0.649	63	1.96
4	0.070	34	0.674	64	2.05
5	0.087	35	0.700	65	2.14
6	0.105	36	0.727	66	2.25
7	0.123	37	0.754	67	2.36
8	0.141	38	0.781	68	2.48
9	0.158	39	0.810	69	2.61
10	0.176	40	0.839	70	2.75
11	0.194	41	0.869	71	2.90
12	0.213	42	0.900	72	3.08
13	0.231	43	0.933	73	3.27
14	0.249	44	0.966	74	3.49
15	0.268	45	1.00	75	3.73
16	0.287	46	1.04	76	4.01
17	0.306	47	1.07	77	4.33
18	0.325	48	1.11	78	4.70
19	0.344	49	1.15	79	5.14
20	0.364	50	1.19	80	5.67
21	0.384	51	1.23	81	6.31
22	0.404	52	1.28	82	7.12
23	0.424	53	1.33	83	8.14
24	0.445	54	1.38	84	9.51
25	0.466	55	1.43	85	11.4
26	0.488	56	1.48	86	14.3
27	0.510	57	1.54	87	19.1
28	0.532	58	1.60	88	28.6
29	0.554	59	1.66	89	57.3
30	0.577	60	1.73	90	---

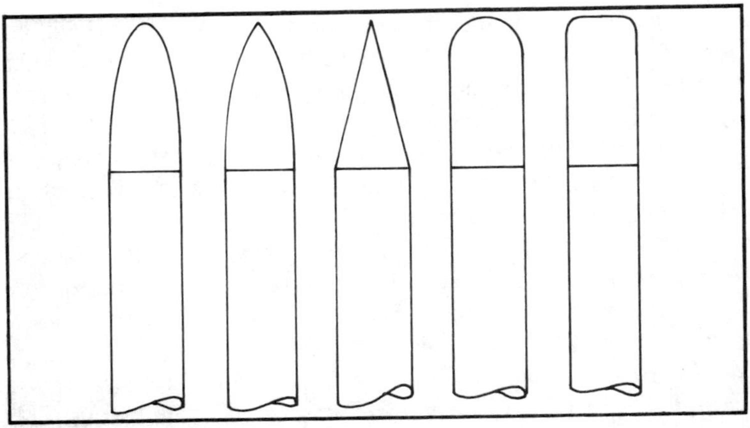

Fig. 6-4. Nose cone shapes.

first glance, it may seem that the sharply pointed design is best, but the sharp point increases drag at subsonic speeds (Fig. 6-5). Thus, the parabolic nose cone is the best possible choice in designing a model rocket for high performance.

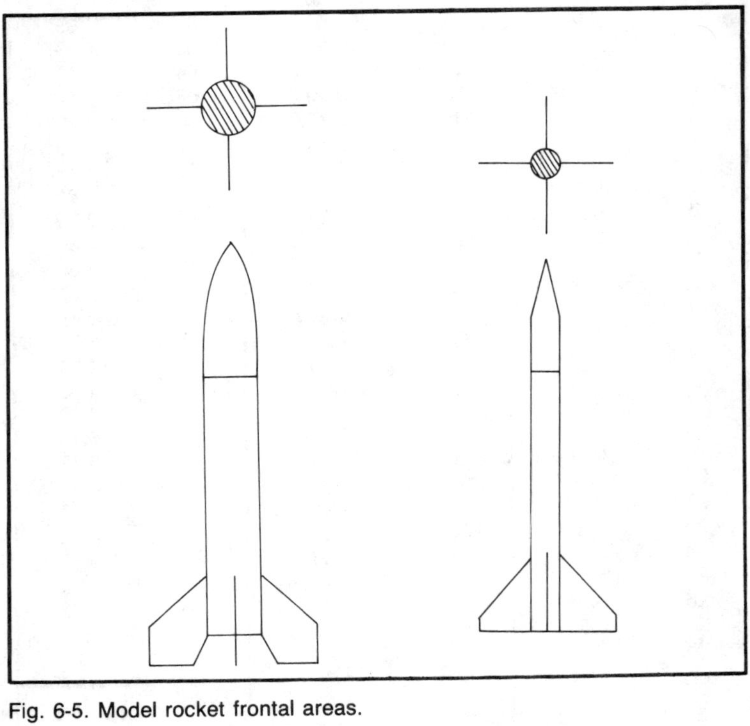

Fig. 6-5. Model rocket frontal areas.

Fig. 6-6. A high drag model rocket. Courtesy Centuri Engineering.

The fuselage of a model rocket is almost always cylindrical, so the shape is not really subject to change in the design stage. The surface of the fuselage can add to the overall drag of a model rocket if it is not smooth, however, because the air flowing past a rough surface can set up vortices and other air flow problems that can increase drag (Fig. 6-6). So, you will want to make the surface of your model rocket' s fuselage as smooth as possible by using proper finishing techniques (careful sanding, the application of several layers of balsa filler coat, and a light coat of paint, smoothly applied).

The cross sectional area of a model rocket has an obvious effect on aerodynamic drag. The larger the area, the more resistance to air flow. Potential resistance here can be overcome by designing with as thin a fuselage diameter as possible (Fig. 6-7).

The base of a model rocket can create a significant amount of drag, as shown in Fig. 6-8. The air passing around the base of the rocket creates a low-pressure area behind the base. This effect can be lessened by rounding the base.

The fins of a model rocket, while necessary for guidance, do create quite a bit of drag. Thus, they should be designed for minimum air resistance. As with nose cones, fins should have rounded leading edges—not sharp or flat. The trailing edges of the fins should have sharp points, as illustrated in Fig. 6-9. For strength, the grain of the wood used to make the fins should be parallel to the leading edges, as in Fig. 6-10. This extra strength is necessary, because you will want to make the fins as thin as possible.

Finally, the fins should be sanded smooth and should have several layers of balsa filler coat and a smooth layer of paint to

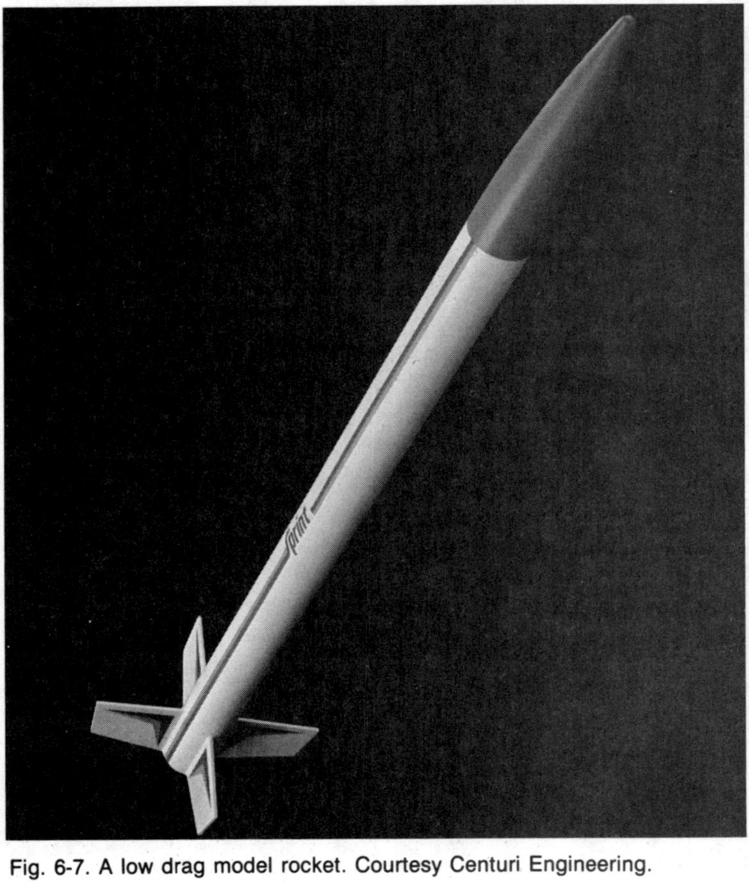

Fig. 6-7. A low drag model rocket. Courtesy Centuri Engineering.

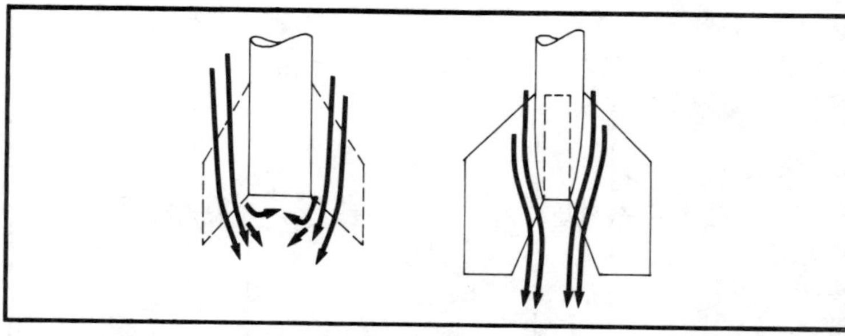

Fig. 6-8. Drag on the base of a model rocket.

reduce surface drag. The nose cone should also be sanded smooth and filled.

The launch lug is a source of quite a bit of drag (up to 15 percent in some models), but it is a necessary evil, unless you use an alternative method of stabilization and guidance during launch, such as the *rail launcher* illustrated in Fig. 6-11.

ALTERNATIVE METHODS OF STABILIZATION

Because model rocket fins do cause a lot of drag, you may want to consider an alternative method of stabilization. Figure 6-12 illustrates one such method, *spin stabilization*. With this technique, smaller fins are required, because the rocket will be spinning like a bullet. The only purpose of the fins in this configuration is to make the rocket spin.

MODEL ROCKET MATH

Much useful information can be obtained from the labels of model rocket engines. As explained in Chapter 1, the label on a model rocket engine provides the total impulse of the engine, and the time delay in seconds between propellant burnout and activation of the ejection charge.

Because the total impulse of a model rocket engine is the total

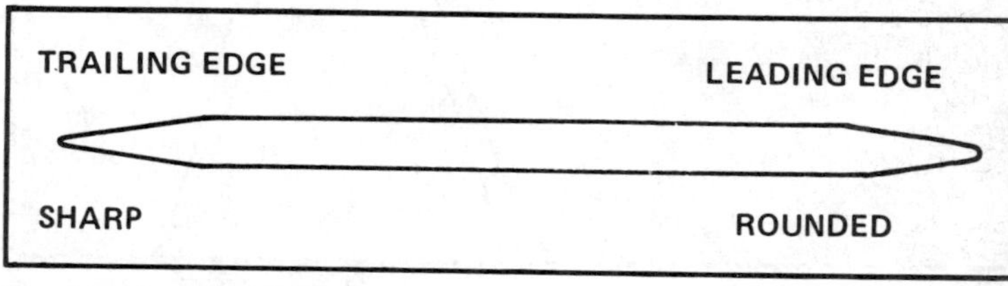

Fig. 6-9. Fin cross section.

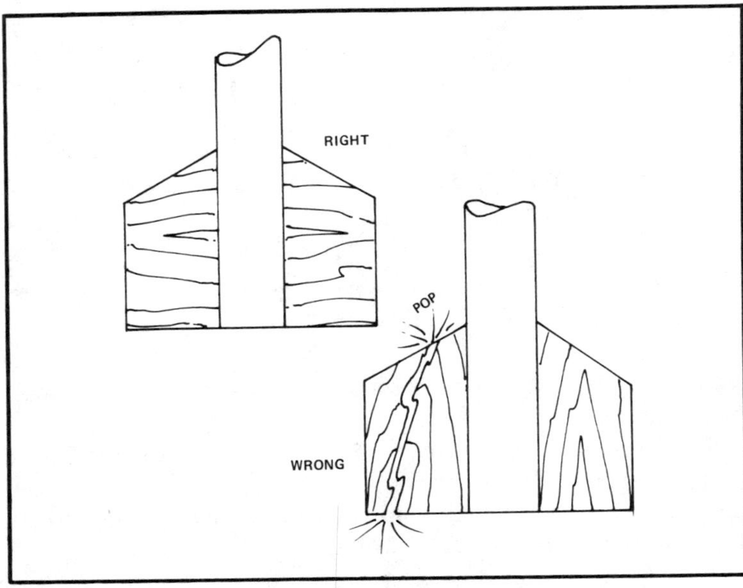

Fig. 6-10. Use proper grain direction when mounting fins.

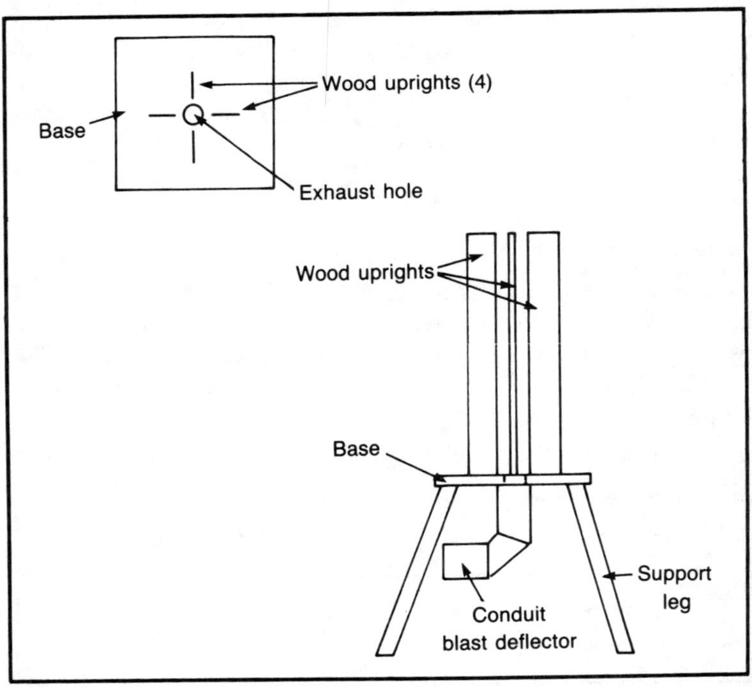

Fig. 6-11. Making a guiderail launcher.

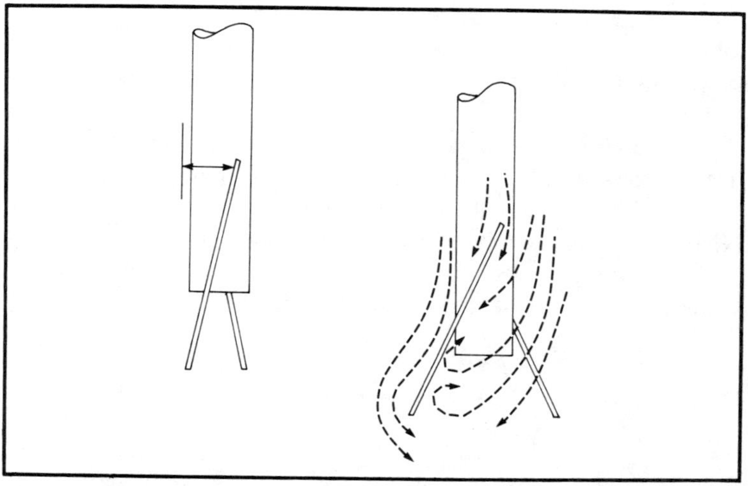

Fig. 6-12. Spin stabilization.

amount of power available (in newton-seconds), and the average impulse is the amount of power available, or being expended, at any second (again in newton-seconds), it follows that the burn time of the propellant can be determined by dividing the total impulse by the average impulse. Thus:

$$t = \frac{I}{F}$$

t = the thrust duration, I = the total impulse, and F = the average impulse.

Thus, the thrust duration of an Estes D12-5 (Fig. 6-13), which has a total impulse of 20 newton-seconds, and an average impulse of 12 newton-seconds, is $\frac{20}{12}$, or 1.7 seconds.

The efficiency of a rocket engine or propellant is described as *specific impulse*. The specific impulse is a measure of the total impulse in pound-seconds that a pound of propellant would deliver. The higher the number, the more efficient the propellant. The formula for determining the specific impulse of an engine is:

$$Isp = \frac{I}{Wp}$$

Isp = specific impulse, I = total impulse, and Wp = the weight of the propellant. The values in this case will be in pounds and pound-seconds, so some conversion will be necessary, unless you have manufacturer's specifications or your own measurement available in pounds.

To convert grams to ounces, divide grams by 28.36. To convert newton-seconds to pound-seconds, divide newton-seconds by 4.46. Next, you must convert the ounces to pounds, which will be a decimal figure, by dividing the number of ounces by 12.

Thus, the pound-second rating of a D12-5 is determined by dividing the newton-second rating by 4.46.

$$\frac{20}{4.46} = 4.48 \text{ pound-seconds}$$

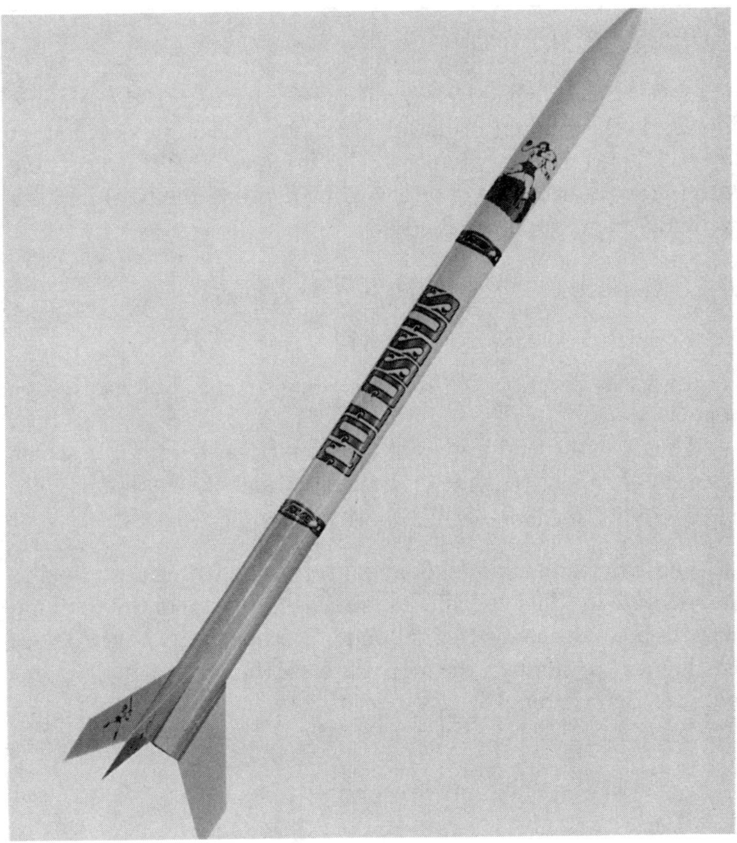

Fig. 6-13. A D-powered model rocket. Courtesy Estes Industries.

The weight of the propellant is determined by first converting the weight in grams to ounces:

$$\frac{24.93}{28.36} = .879 \text{ ounces}$$

and then converting the ounces to pounds:

$$\frac{.879}{12} = 0.73 \text{ pounds}$$

Now, you can determine the specific impulse of the D12-5 engine by using the previously mentioned formula, $Isp = \frac{I}{Wp}$. Simply divide the total impulse in pound-seconds (4.48) by the weight of the propellant in pounds (.073) to arrive at the specific impulse figure of 63.

An engine's specific impulse can be used to determine the *exhaust velocity* of the engine (the speed at which the expanding gasses are exhausted through the engine's nozzle). The formula for determining exhaust velocity is:

$$C = G \times Isp$$

C = the exhaust velocity, G = the *force of gravity* (32 feet per second per second), and Isp = the specific impulse.

Looking again at the D12-5 engine from Estes, it is possible to determine that it has an exhaust velocity of 2016 feet per second:

$$(32 \times 63 = 2016)$$

Exhaust velocities of various engines are graphic examples of why a higher specific impulse is more efficient than a low one. *Newton's Third Law* states that for every action there is an equal and opposite reaction; obviously, the greater the action, the greater the reaction. In terms of exhaust velocity, there will be a larger reaction (movement) on the part of a rocket with an exhaust velocity of 2016 feet per second, than one with a velocity of 640 feet per second, which happens to be the exhaust velocity that an engine with a specific impulse of 20 would yield.

Thus, engines with a high specific impulse (which means that they have a high total impulse and low propellant weight) provide a greater "kick," or reaction, at lift off than do those with a low specific propellant weight. In the same manner, engines with a high average impulse and short burn time will provide more of a "kick" at lift off than those with a low averge impulse and long burn time (or *thrust duration*).

Knowing the performance parameters of model rocket engines, such as those just described, can help you in determining which engines to use to power your own designs. If you have a heavy rocket, and/or one with a good deal of aerodynamic drag, you'll want to choose an engine that gives it a good "kick" at lift off—in other words, one with a high specific impulse, or a high total impulse combined with a high average impulse.

If you desire a long burn time, you must have a lightweight rocket, because a lighter rocket doesn't require that extra "kick" at liftoff.

Another factor that enters into the design of a model rocket is the *mass ratio*. The mass ratio is the relationship between the takeoff weight of a rocket and the weight of the rocket after the propellant has been expended. This is determined by dividing the initial weight of the rocket by its burnout weight, thus:

$$n = \frac{Wi}{Wi - Wp}$$

n = the mass ratio, Wi = the initial weight of the rocket, and Wp = the weight of the propellant.

If all other conditions are the same, a rocket with a high mass ratio will reach a greater velocity than one with a low mass ratio, because a greater amount of the weight of the rocket with the high mass ratio is propellant, which will be exhausted as the propellant burns.

Using this formula, you can determine that the mass ratio of the Estes D12-5 in a rocket that weights 5 ounces is 1.21 (all figures are in ounces, and the initial weight includes the weight of the engine).

If you substitute an Estes D11-9 engine, the mass ratio lowers slightly, to 1.14. This is because the initial weight of the D11- 9 is .01 ounce less than the initial weight of the D12-5, which makes the initial weight of the rocket 4.99 ounces:

$$\frac{4.99}{4.99 - .65} = 1.14.$$

(Note, too, that the weight of the propellant is somewhat less than that of the D12-5.)

(It is possible to determine any missing factor in any of these equations by transposing the equation in question.)

Acceleration and speed can be determined several different ways. If you know the speed of an object and how long it took to achieve that speed, you can divide the total speed by the time to obtain the average acceleration. Thus, if a model rocket accelerated for 7 seconds to attain a speed of 210 mph, its average acceleration was 30 mph *per second*, or;

$$210 \div 7 = 30$$

If you know the rate of acceleration, you can determine the final speed by multiplying the acceleration by the time. If the acceleration is constant, as with the speed of an automobile, things are fairly simple. With a model rocket, however, you have a changing rate of acceleration.

To determine the average rate of acceleration, you must first calculate the *G-force* (multiples of the force of gravity). The G-force on a rocket in space is determined by dividing the thrust of the engine by the weight of the rocket:

$$G = \frac{F}{W}$$

G = the G-force, F = the thrust of the engine, and W = the weight of the rocket. A 6-ounce rocket being subjected to a thrust of 10 ounces is subject to a force of 1.6 G's. Thus, a payload in the rocket weighing 1 ounce would weigh 1.6 ounces. If the thrust were altered to 20 ounces, the G-force would change to 3.33, and a payload would be subjected to over three times its weight.

Knowing the G-force on a rocket, you can next determine how much speed it will gain in a given length of time (acceleration). A "G," which is equal to the force of the Earth's gravity, is equal to an acceleration of 32 feet per second (that's 32 feet per second increase in speed *each* second).

Because model rockets are being launched from the surface of the Earth, you must take into consideration that 1 G of acceleration is being expended to overcome the force of gravity. So, the acceleration of a model rocket is actually the thrust divided by the weight minus 1 G. Thus:

$$G = \left(\frac{F}{W} - 1\right)$$

Once you find the G-force acting on a rocket, you multiply this figure by 32 (the rate of acceleration of gravity) to find the rocket's rate of acceleration, *per second*, using this formula:

$$Ao = \left(\frac{F}{W} - 1\right) \times 32$$

If you multiply this figure, which is the rate of acceleration, by the duration in seconds of the acceleration, you will have the final velocity, or speed.

For example, if you subject a 6-ounce rocket to 10 ounces of thrust for 5 seconds, this is the result:

$$Ao = \left(\frac{10}{6} - 1\right) \times 32 \times 5 = 106.56 \text{ feet per second.}$$

A heavier rocket, say one which weighs 8 ounces, will naturally have a lower final speed, because it will be subjected to a lower G-force.

These figures can also determine the approximate altitude that a rocket reaches, because distance is equal to average velocity multiplied by the duration. A rocket traveling 100 feet per second for 3 seconds will reach an altitude of 300 feet, provided the speed is constant. Unfortunately, it is not constant, so you must determine an average velocity for the rocket. This is done by breaking the flight down into intervals and averaging the speed by intervals.

To do this, measure the velocity at two points during the flight and divide the sum of the two velocities by 2. If you measure the velocity at liftoff as zero and add this to the final velocity of 106.56 feet per second, you have an average velocity (after dividing by 2) of 53.28 feet per second. Multiplying this average velocity by the duration of 3 seconds, you have a final altitude of 159.84 feet. The formula for determining this altitude at burnout is:

$$H = \left(\frac{V'' + V}{2} \right) \times t$$

H = altitude, V″ = velocity at first interval, V = velocity at second interval, and t = duration.

The rocket will, of course, coast a bit farther after propellant burnout, slowing down as it coasts because there is no thrust to maintain acceleration.

In the absence of thrust, the rocket is subject to the same force of gravity that it was during acceleration, which is 32 feet per second per second, as stated previously. Thus, it will lose 32 feet per second of its velocity each second.

With this knowledge, it is possible to predict how long your hypothetical rocket will coast after propellant burnout. You simply divide the rocket's burnout velocity (106.56 feet per second) by 32 (the acceleration of gravity) to find the number of seconds it will coast. In this case, the answer is 3.33 seconds.

To find how far the rocket will coast after burnout, use the formula for determining altitude:

$$H = \left(\frac{V'' + V}{2} \right) \times t$$

Fig. 6-14. High performance rocket in flight.

In this instance, the rocket would coast upward 177.42 feet.

Thus, the total predicted altitude for your rocket would be 337.26 feet (Fig. 6-14).

This, of course, is assuming that the recovery system does not deploy before the rocket completes coasting to apogee and that there will be no significant effects from aerodynamic drag.

It is possible to use predictions such as this one to determine just how long a delay time must be used—in this case, it should be over 3 seconds. Drag increases as the velocity increases (in fact, it increases four times as velocity is doubled). The calculation of drag is a very complex subject and there have been volumes written on the subject. If you are interested in pursuing the topic further, check the manufacturers' catalogs for appropriate titles.

Chapter 7

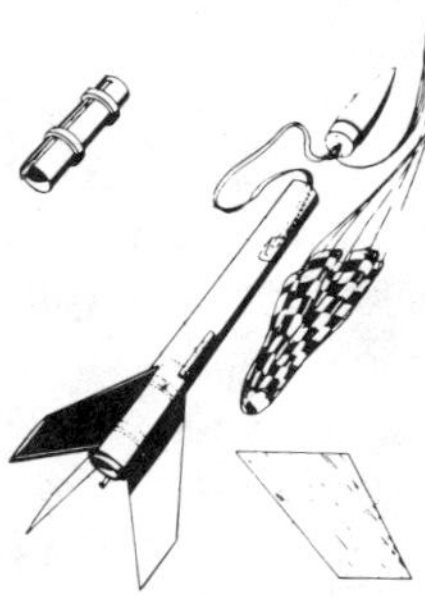

Payloads, Research and Development, and Advance Design

Almost every rocketeer will, sooner or later, decide that it might be interesting to fly a rocket with a small animal as a payload. As attractive as this idea may be, it really serves no useful purpose. Small creatures such as mice will almost invariably be killed by the acceleration of a model rocket (provided the rocket is designed with enough power to get off the ground). This will tell you nothing that you cannot determine by reading books such as this one. Fish are somewhat better able to take high acceleration if they are inside a container full of water during the flight, but this will probably put your rocket over the weight limit of 1 pound.

Insects, on the other hand, make ideal "astronauts." High acceleration cannot harm them because they have exoskeletons, and they make good subjects in that they can be observed for changes in habit after a flight. A grasshopper may, for example, be quite disoriented after a flight in a model rocket and may be unable to fly for very great distances. A useful experiment would be to fly several grasshoppers as passengers in model rockets and observe their behavior both before and after their flights. The purpose of the experiment would be to determine if there are any common effects experienced by the grasshoppers.

INSTRUMENTATION

Useful data can be gathered during a model rocket flight for the purposes of determining the accuracy of a prediction or for gaining more information about model rocket performance or atmospheric conditions at high altitudes.

Figure 7-1 will show you how to construct a mechanical *accelerometer.* The accelerometer will tell you how many G-forces your model rocket was subjected to, and from this you can determine the acceleration of the rocket or verify your prediction of the rocket's acceleration, using the formulas in Chapter 6.

For high altitude model rockets and those carrying expensive or important payloads, a *locator beacon* is vital. An audio beacon can be constructed following the plans in Fig. 7-2, or one may be purchased from CNA (see address in Appendix A). For night flights, you may wish to add a visual beacon to your model rocket. This is simply done by creating a transparent payload section and placing a light in it, as in Fig. 7-3.

Telemetry devices carried onboard model rockets are extremely useful for recording on-the-spot information for later analysis. The

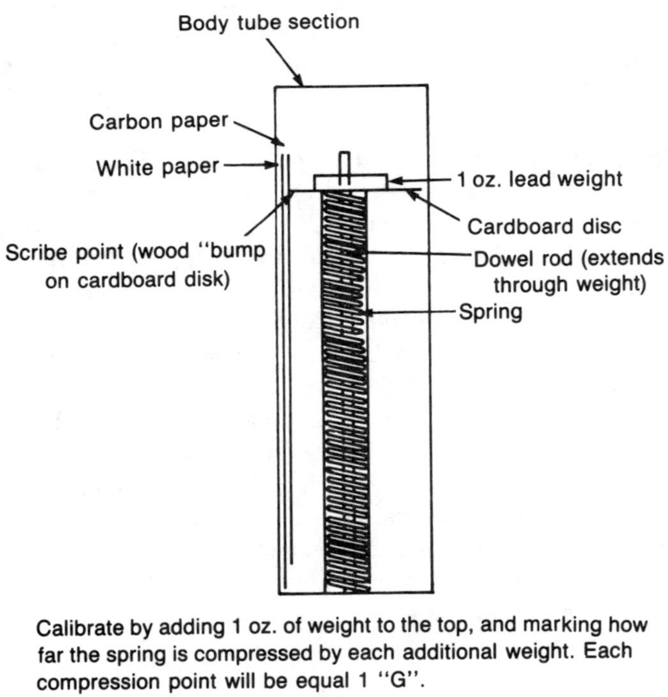

Calibrate by adding 1 oz. of weight to the top, and marking how far the spring is compressed by each additional weight. Each compression point will be equal 1 "G".

Fig. 7-1. Making a mechanical accelerometer.

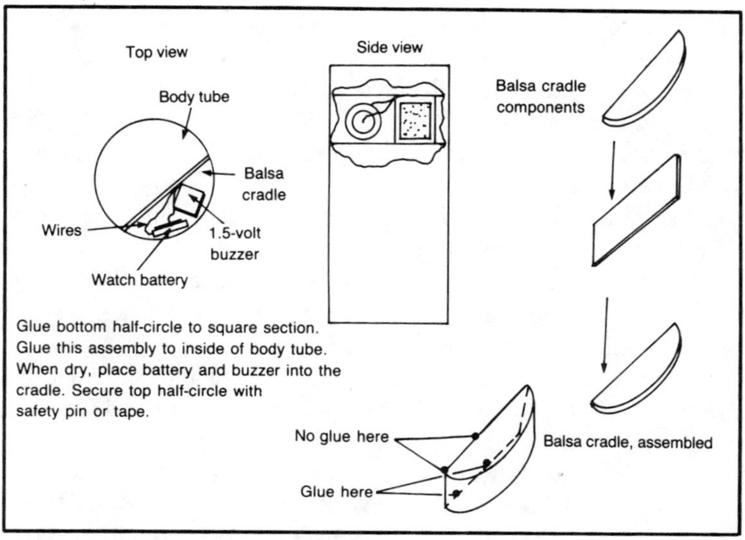

Fig. 7-2. An audio recovery beacon.

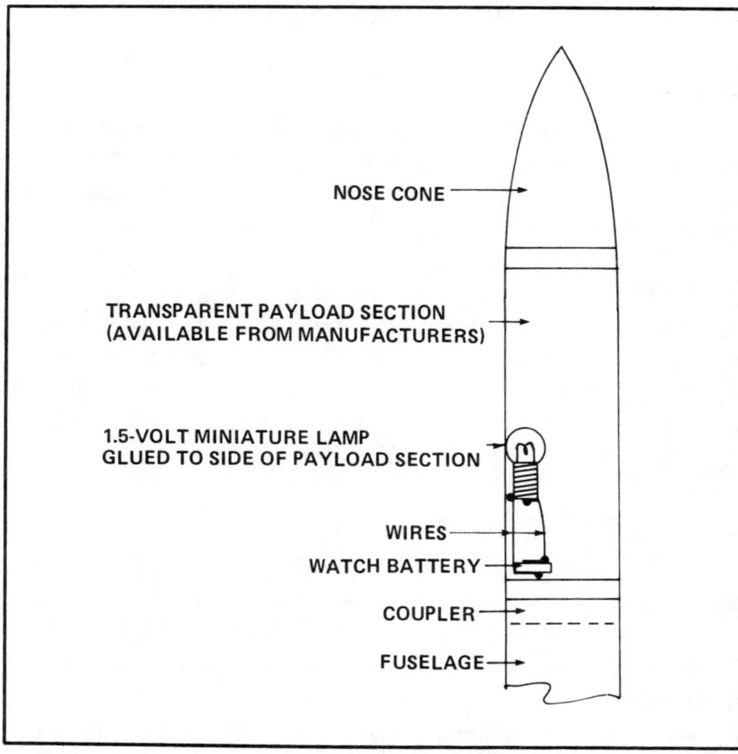

Fig. 7-3. Transparent payload section with light.

Transrod, manufactured by Estes Industries until the early 1970s, was a fine example of miniature technology. Weighing only a few ounces, the Transroc was designed to fit into a BT-50 body tube and could transmit data on any of 23 CB channels. With various modifications, it could act as a radio locator beacon, a microphone, a light/spin rate sensor, or an accelerometer. Though no longer being manufactured, the Transroc can still be found in some hobby shops and is worth looking for.

The successor to the Transroc is the Foxmitter, available from Competition Model Rockets. It has many of the same performance characteristics of the Tarnsroc.

If you are experienced in electronics design and construction, you may find it easy to make your own version of the Transroc. (Be sure to stay under the weight limit!)

The information from a microphone, transmitted back to a ground- based recorder, can provide you with an accurate time-frame for flight events such as propellant burnout, activation of the ejection charge, and touchdown. A light sensor can provide you with information on the spin rate of the rocket.

CONTEST PAYLOADS

The standard National Association of Rocketry payload is a 1- ounce lead weight. (This is in the process of being changed to a more frangible sandbag.) Many competition events require the use of this standard payload to add to the difficulty of winning altitude events.

Another type of contest payload is a raw egg. Several events, including altitude competition and parachute duration, can be flown with an egg as a payload (Fig. 7-4). The extra challenge of returning the egg to earth unbroken makes this a run event, as well as a tribute to the design abilities of those whose eggs do return unbroken.

AERIAL PHOTOGRAPHY

Aerial photography—still and motion—is probably the most popular of model rocketry activities, beyond simply flying model rockets. Soon after the beginnings of the hobby in the late 1950s, model rocketeers were attempting to fly various types of cameras aboard model rockets.

The various experimenters met with some success, but only after quite a bit of trial and error. It was not until the mid- 1960s,

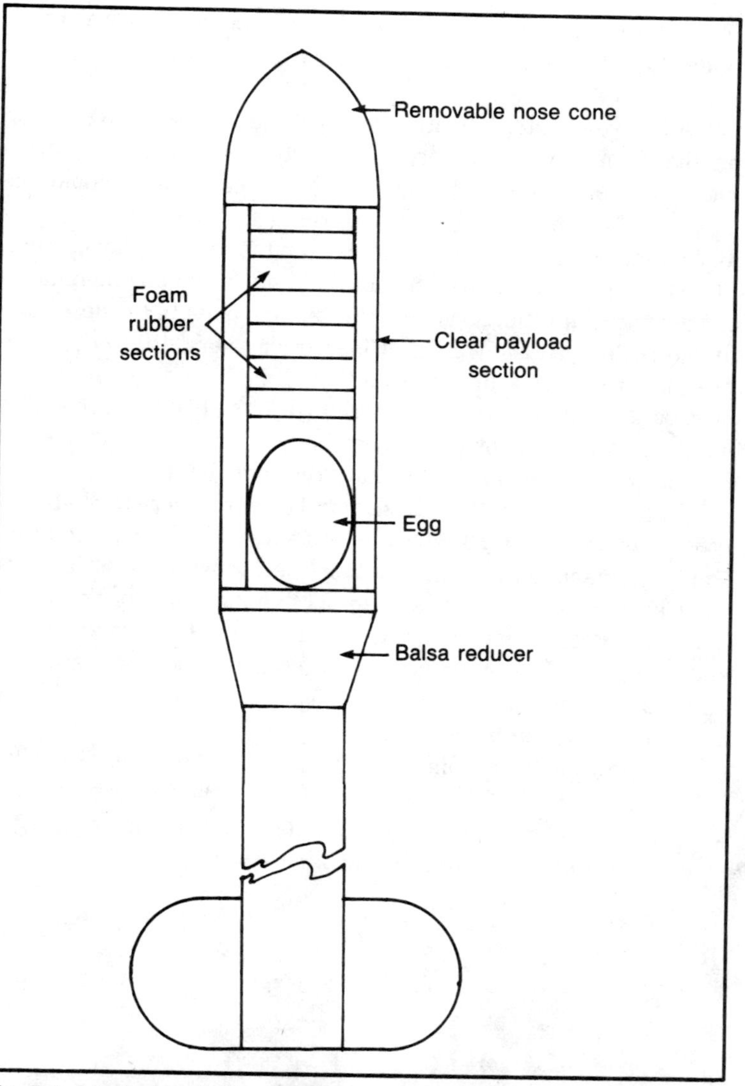

Fig. 7-4. A typical egglofter design.

when Estes introduced the Camroc, that aerial photography became a wide popular activity (Fig. 7-5).

The Camroc was a small cylinder, as shown in the accompanying photo, which was loaded with a 1 1/2-inch diameter circular negative of Tri-X black and white film. Flown with an extra long delay between propellant burnout and ejection, the Camroc and its carrier rocket tilted over at apogee and began descending before the picture was taken. The lens of the Camroc faced forward and, at ejection, the entire assembly was ejected, permitting the shutter to snap.

A few years after the introduction of the Camroc, Estes scored another first with the Cineroc. The Cineroc was a miniature motion picture camera that used Super 8mm film in cartridges. Unlike the Camroc, the Cineroc (Fig. 7-6) exposed its film— approximately 40 seconds' worth in semislow motion—on the way up. Thus, a complete film record could be made of the flight, from engine ignition though parachute ejection, from the vantage point of the nose cone of the rocket. As shown, the Cineroc was very different in appearance from the Camroc, even though both cameras are in the nose cones of their respective carrier rockets. The Cineroc lens is mounted in the side of the camera body, and a mirror is used to convey the image from the ground through the lens.

Unfortunately, both of these aerial cameras were taken out of production in the early 1970s, due to a shortage of parts and problems with manufacturing molds. Like the Transroc, both Camrocs and Cinerocs can sometimes be found in hobby shop back-stocks.

There is an aerial camera currently available for model rockets, however. The Astrocam, another innovation of Estes Industries, was introduced in 1979 and has proven itself to be a highly useful piece of equipment (Fig. 7-7).

As shown in the accompanying photo, the Astrocam is similar in appearance to the Cineroc (Fig. 7-8). Like the Cineroc, it takes a photo looking through a side-mounted lens into a mirror that provides a view of the ground below. The Astrocam's photo-taking

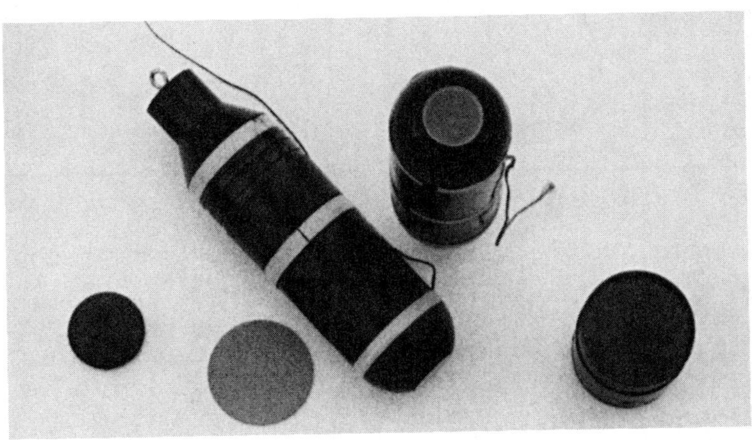

Fig. 7-5. The Camroc.

Fig. 7-6. The Cineroc.

technique more closely resembles that of the Camroc, however.

Like the Camroc, the Astrocam coasts to apogee and, thanks to an extra-long delay charge, turns over and begins to descend before the ejection charge is activated. The shutter is activated by the release of a string when the Astrocam is ejected. The release of the string releases the tension on a spring and this snaps the camera's shutter across the lens opening. One shot is made per flight.

The Astrocam uses 110 film cartridges. 400 speed film is recommended because of the fast shutter speed (1/500th of a second) required to "freeze" the picture. (The camera is moving downward at a high rate of speed when the shutter is activated, so a slower film would result in a blurred picture and possible underexposure.) The focal length of the camera is 30 mm at f/16.

Fig. 7-7. The Astrocam. Courtesy Estes Industries.

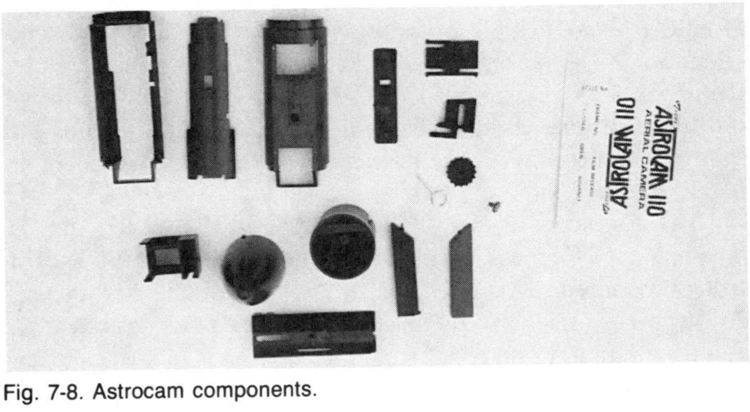

Fig. 7-8. Astrocam components.

The Astrocam is sold with its own carrier rocket, the Delta II. The Delta II is powered by B or C engines, and Estes offers a D-powered launch vehicle for the Astrocam as well (Fig. 7-9). (Those interested in constructing their own launch vehicles for the Astrocam will find that none of Estes' body tubes fit the camera body. Centuri Engineering Company's "Series 20" body tube, however, is the exact size needed.)

As stated before, an extra long delay is required when flying the Astrocam, to allow the rocket to tilt over and begin its descent before the shutter is activated by ejection. Estes recommends B8-5, B8-7, and C6-5, and C6-7 engines for low- and medium-altitude

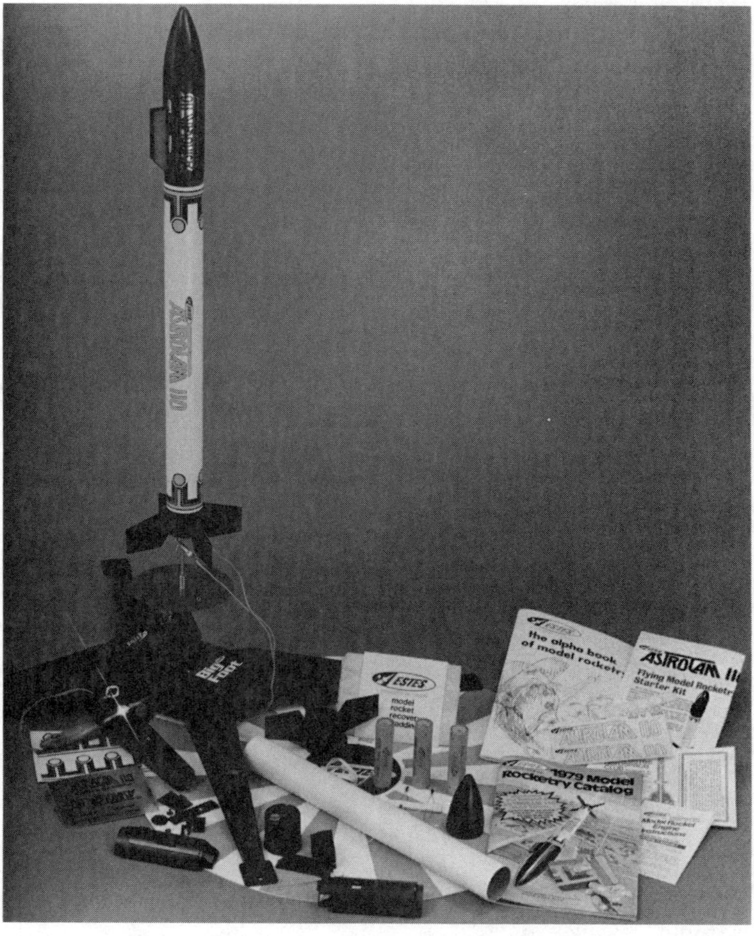

Fig. 7-9. The Astrocam starter kit. Courtesy Estes Industries.

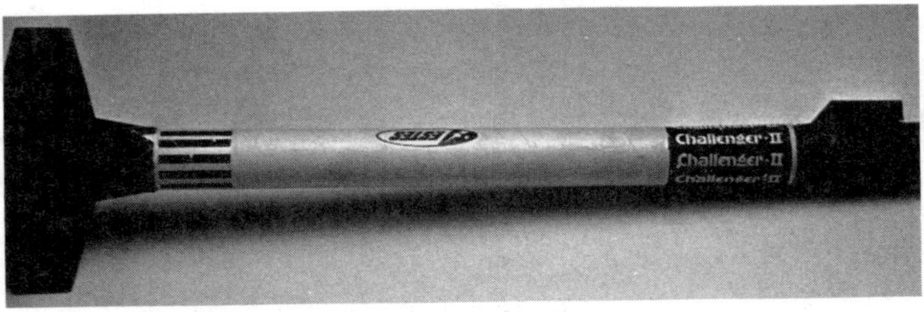

Fig. 7-10. The Challenger II by Estes.

flights, but a C6-3 engine will work as well for horizon shots. For high-altitude flights, Estes recommends their D11-9 engine, but the D12-7 works just as well (Fig. 7-10).

It is helpful to carry a checklist, such as that shown below, and refer to it during each and every Astrocam launch. This may seem finicky, but anyone who has used Astrocam can attest to the number of shots missed due to not setting the shutter or leaving the light-lock closed. The checklist steps are as follows:

1. Film in Camera.
2. Astrocam assembly secure (access panel in place).
3. Engine placed in carrier rocket, with igniter in engine.
4. Film advanced.
5. Shutter cocked and string secured between base of camera and fuselage.
6. Astrocam rotated so that optics hood is opposite launch lug(s) on fuselage.
7. Astrocam and carrier placed on launch pad.
8. Igniter leads attached.
9. Light lock open!
10. Continuity of launch controller circuit checked.
11. Launch range clear.
12. Recovery crew in place.
13. Flight path clear of low-flying aircraft.
14. Countdown.
15. Launch!

If you follow this checklist for every Astrocam flight, you should have almost no problems with getting photos and recovering your rocket safely. Some of the items on the checklist may seem petty, but it's easy to miss one and thus foul up a flight or waste a flight without getting a photo.

After you have flown the Astrocam several times, you may be

interested in trying to take photos looking back over the carrier rocket's fins, on the way up. This is called the *reverse Astrocam configuration* and can be constructed using the plans in Fig. 7-11. When flying the Astrocam in this configuration, you will want to use a short delay time (3 to 5 seconds) so that the camera will be looking at the ground at ejection (Figs. 7-12 and 7-13).

If you have more than one Astrocam, it is possible to fly two or more strapped on to a large rocket, as shown in Fig. 7-14. Two Astrocams placed close together will yield what is known as a *stereo*

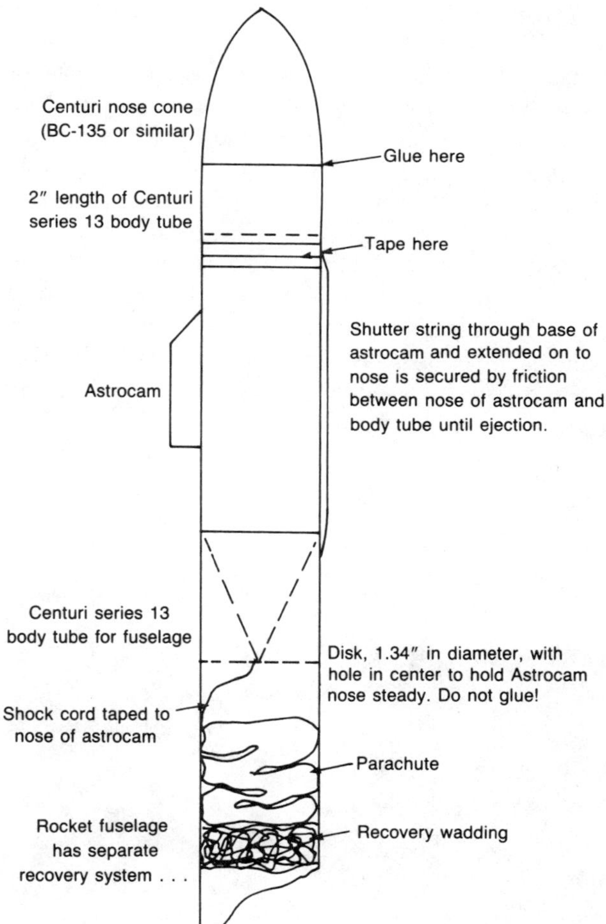

Fig. 7-11. A reverse Astrocam plan.

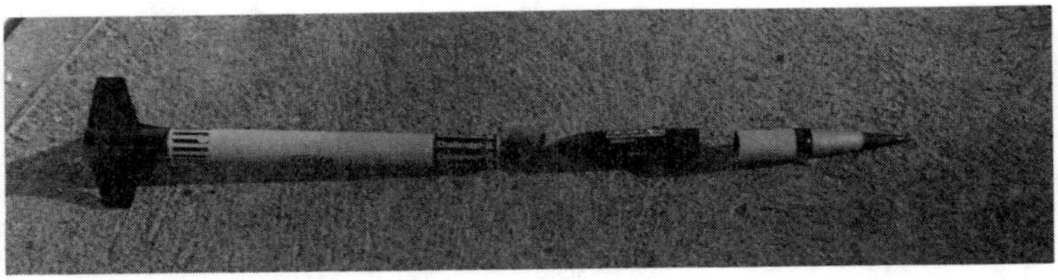

Fig. 7-12. Reverse Astrocam components.

pair. The stereo pair, when viewed close up with a strip of cardboard separating them, will give a three-dimensional view of the area photographed.

You may be curious as to what sort of information can be gained from aerial photographs. Surprisingly enough, much valuable information can be obtained. For example, fields of corn or other crops, when viewed from a high altitude, look markedly different than when viewed from ground level, and areas of diseased plants, or poorly growing plants, can be easily distinguished. In the same manner, it is possible to spot schools of fish and varying depths in large bodies of water by color differences.

Old roads and building foundations can also be outlined from an aerial view when they are nearly invisible from the ground, and geological features are highlighted in an aerial view.

Using a simple formula, it is possible to determine the size of objects shown in a photo and to find the altitude of the camera when a photo was made. The formula is:

$$H = \frac{OF}{I}$$

H = the altitude of the camera, O = the size of an object on the ground in the photo, F = the focal length of the camera, and I =

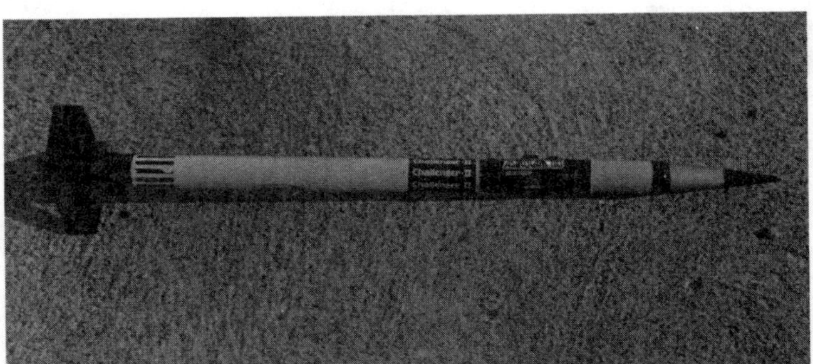

Fig. 7-13. Reverse Astrocam.

Fig. 7-14. Strap-on parasite Astrocams.

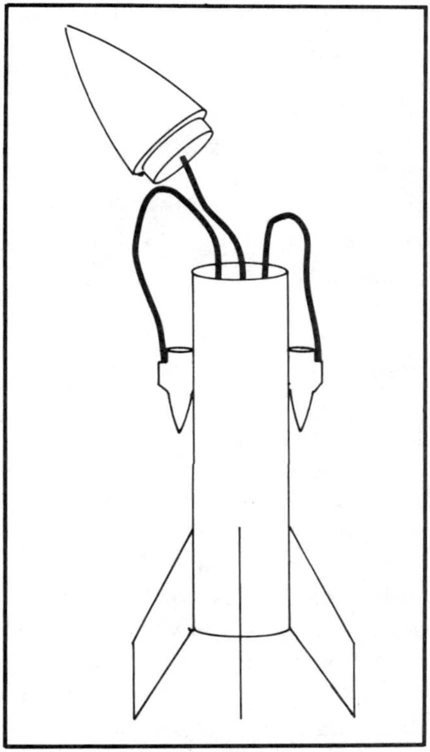

the size of the object's image as measured on the negative in millimeters. (Millimeters are used in the measurement on the negative because the camera's focal length is given in millimeters, and small measurements of this type are more easily made in millimeters.) The other distances will be in meters.

Because the focal length of the Astrocam is a constant 30 mm, the formula for determining altitude can be expressed as:

$$H = \frac{O \times 30}{I}$$

In the case of a wall that is 10 meters long on the ground, and which measures 5 millimeters on the negative, you have:

$$H = \frac{10 \times 30}{5} = 60$$

The camera was 60 meters above the wall when the photo was made.

To use the formula to determine the size of an object shown in the photo, transpose it to:

$$O = \frac{HI}{30}$$

Thus, if the camera was at an altitude of 60 meters when the photo was made, and the wall measures 5 millimeters on the negative, you have:

$$O = \frac{60 \times 5}{30} = 10$$

The size of the wall is 10 meters.

A few additional considerations when flying the Astrocam include the size of the recovery field, the wind, and the aiming of the Astrocam.

Obviously, because the Astrocam is a little more costly than most model rockets, you'll want to make sure that you have a large recovery area. If you want to photograph a particular site that is not surrounded by a large field, however, you should use streamer recovery, or you may want to consider recovering the carrier rocket via streamer and the camera itself with a very small parachute.

The wind will affect the flight and recovery of the Astrocam just as it does any rocket, although the Astrocam is heavier than many model rockets and will tend to drop more quickly. Again, the Astrocam is something that you don't want to lose over the horizon or to a tree, so take extra caution when flying it (refer to Chapter 4 for additional flying tips).

It is possible to "aim" the Astrocam to a limited extent. By tilting the launch rod, you can at least get the Astrocam into the general area you wish to photograph. Obtaining a photograph of a particular building or other site will most likely require several flights, however.

Despite its limitations, the Astrocam is the most economical method of taking aerial photographs now available, and it adds yet another dimension to the hobby. Future developments in the area of aerial photography with model rockets may well include a revival of motion picture photography and multiple exposure still photography. Model rocket manufacturers are currently in the process of developing such cameras, but only time will tell if they will reach the market.

ONBOARD COMPUTERS

Thanks to the miniature electronics technology brought on by the space program, it is possible to make very small, ultra-lightweight electronic devices for model rocket payloads. Dedicated electronics buffs have succeeded in flying onboard computers that take air samples; record performance parameters such as acceleration, thrust duration, and flight time; and monitor the status of other payloads while a model rocket is in flight. The most notable efforts in this area have been by Charles Hall, Jr., a physics student at Ohio State University. In addition to collecting and storing information such as that listed above, Hall's computer "dumped" its data into a ground-based computer that provided interpretation of the rocket's flight. An experimental effort, Hall's computer tipped the weight of its carrier rocket just over the legal limit, but new advances in large scale integration of electronic circuits will soon make it possible to build more elaborate onboard computers for model rockets that will stay well within the prescribed weight limits.

STATIC TESTING OF MODEL ROCKET ENGINES

For experimental purposes, or to satisfy your own curiosity, you may wish to measure the thrust of various model rocket engines. There is very expensive electronic test equipment available for this purpose, but you don't have to buy it. You can build your own *test stand* quite inexpensively, using some very common components.

Figure 7-15 provides the basic plans for building your own test stand. When static testing a model rocket engine, be sure to fasten the engine down very securely, use a heavy base for the stand, and provide a blast deflector to channel the engine's exhaust safely.

WIND TUNNELS

Aircraft and missile use *wind tunnels* to test stability and performance characteristics of their designs in the form of models before they are put into production. You can do the same with your model rockets.

A simple wind tunnel can be constructed, with readily available components, using the plans in Figs. 7-16, 7-17, and 7-18. Observing the performance of your model rockets in the wind tunnel will not only tell you a lot about their potential stability, but it will also illustrate the effect of air movement over and around the nose cone, fuselage, and fins of the rocket.

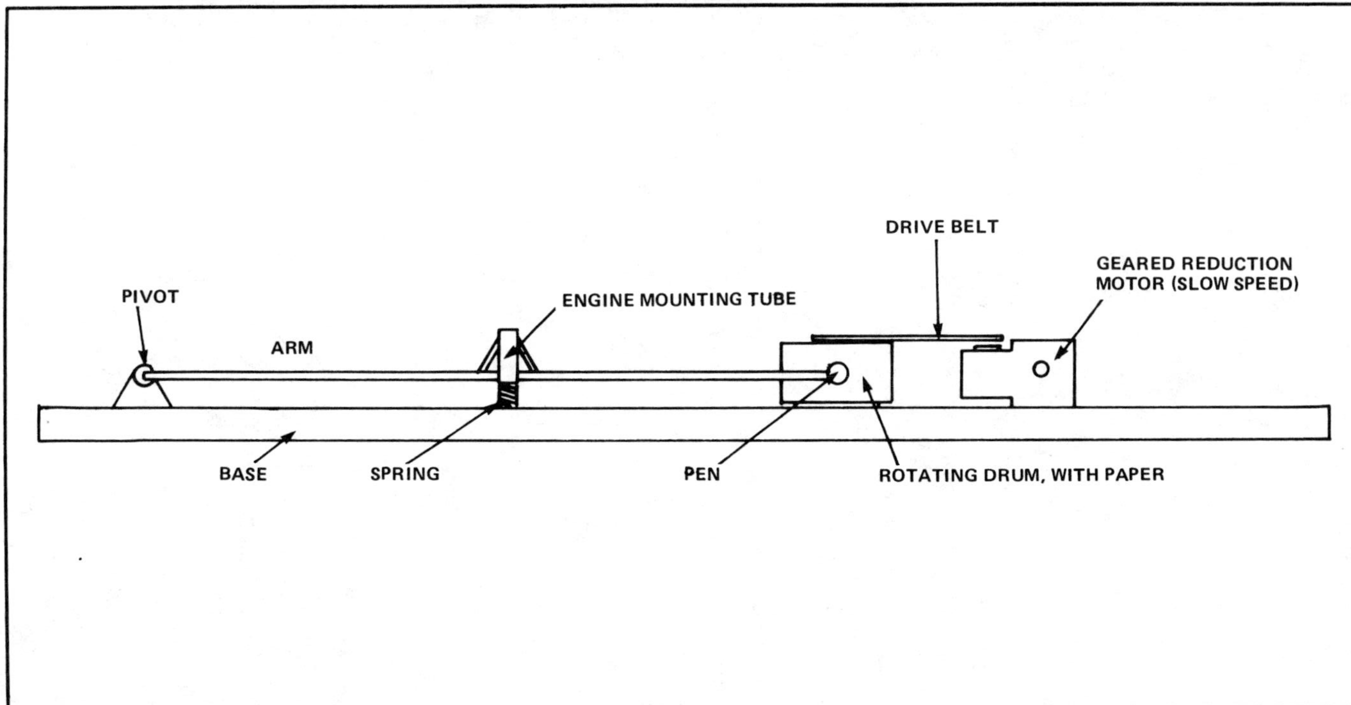

Fig. 7-15. An engine test stand.

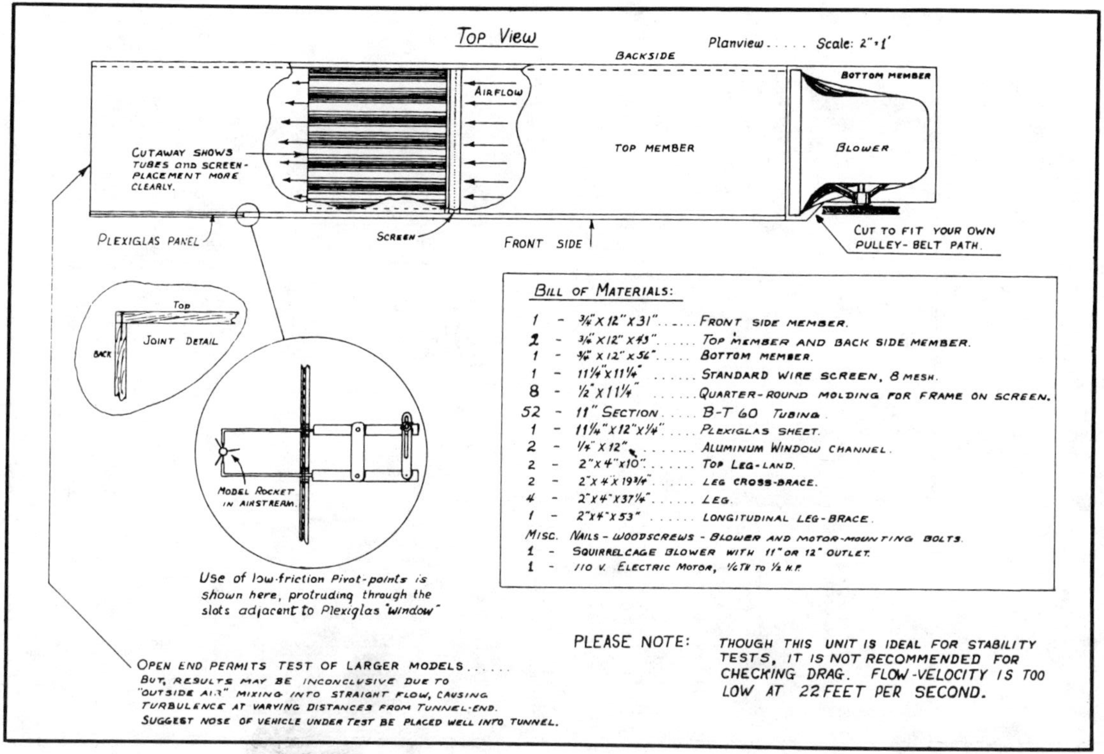

Fig. 7-16. Wind tunnel plan, top view. Courtesy Estes Industries.

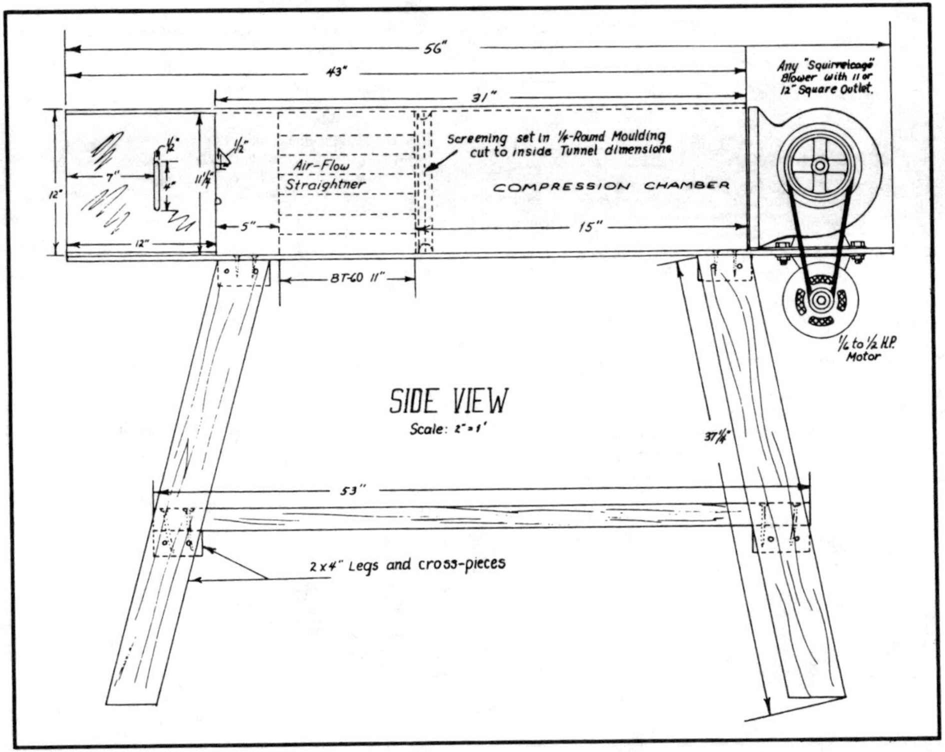

Fig. 7-17. Wind tunnel plan, side view. Courtesy Estes Industries.

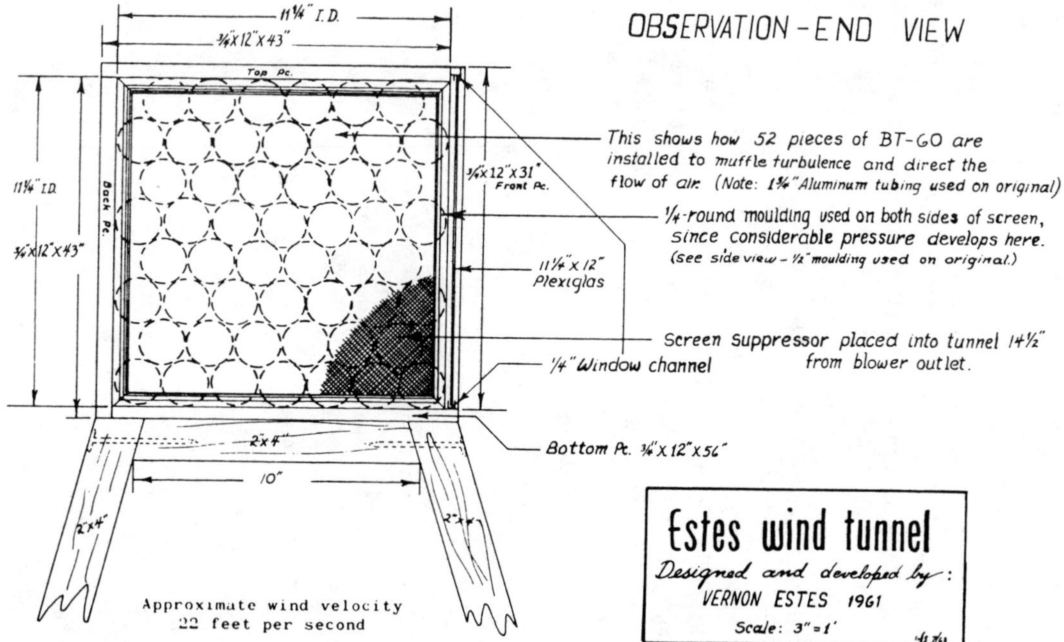

Fig. 7-18. Wind tunnel plan, observation end view. Courtesy Estes Industries.

EXPERIMENTAL RECOVERY SYSTEMS

Much experimentation has been done in recent years with model rocket recovery systems, because the recovery system is the "weakest link" in the model rocket flight profile. During recovery, the rocket is usually at the mercy of vagrant winds, and more flights are ruined by no recovery or unplanned landings on roofs or in trees than by any other cause.

Time delay of parachute ejection beyond that afforded by the model rocket engine's delay charge has been an area of much concentration. Radio control of parachute ejection offers the most promise in this area. The usual approach is similar to that shown in Fig.7-19, where a small *drogue parachute* or streamer is ejected by the engine's ejection charge. The drogue is attached via string to the main parachute, which in turn is restrained by a string. The string restraining the main parachute passes over a flashbulb. When the rocket is close enough to the ground, the rocketeer can activate a switch, via radio control, which completes a circuit between the flashbulb and a battery. The flashbulb is ignited and burns through the restraining string. This releases the main parachute, which is pulled out by the drogue.

Nose cone is anchored to a ring set in the body tube. The main chute's shock cord is also anchored to this; it will not deploy until drogue chute is released.

Drogue chute is deployed by the ejection charge, as it is inserted in the internal stuffer tube. The line from the drogue chute is anchored to the release mechanism, which prevents its pulling the main chute out of the body tube until the timer activates the release mechanism. After a preset time, line holding back the drogue chute is released and the main chute is deployed.

The timer can be started manually, or with a sensor to detect the engine's ignition.

Main chute should be deployed when the rocket is 200-300 feet above the recovery area.

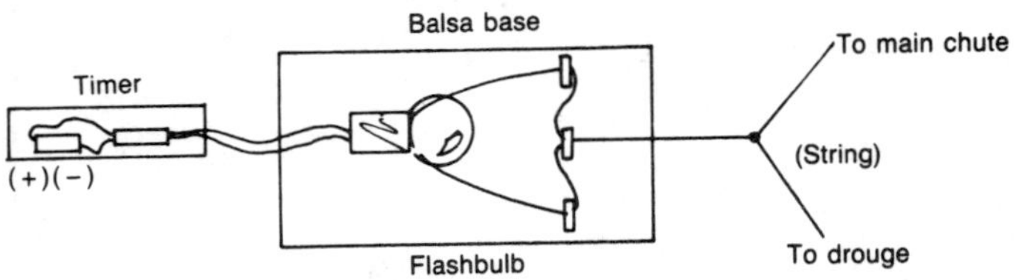

Fig. 7-19. Drogue chute and parachute delay system. Courtesy *Model Rocketeer Magazine*.

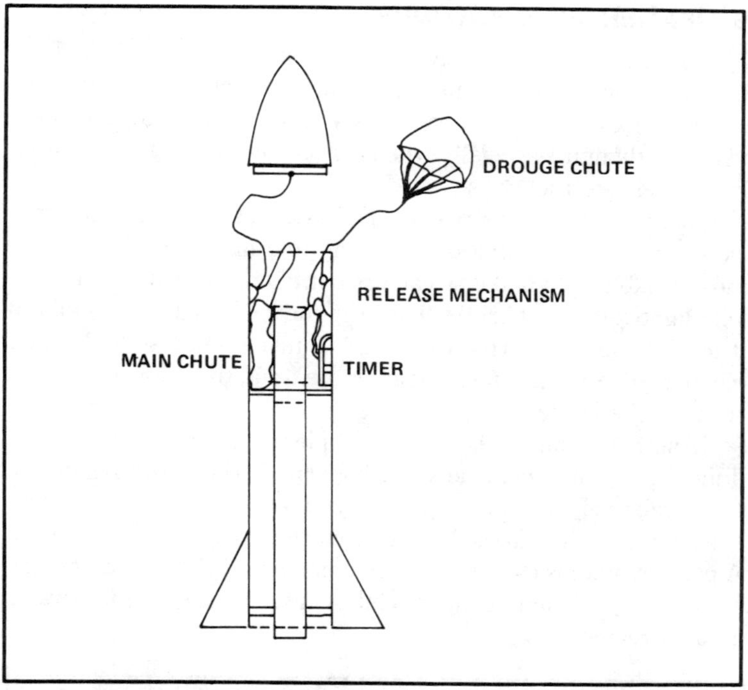

Fig. 7-20. A parachute delay timer system.

Using this method is a matter of trial and error, because one must use his own judgment as to when it is safe to release the main parachute.

Similar to the radio control method is the technique for parachute delay illustrated in Fig. 7-20. Here, a timer is used in place of radio control to close the circuit between the flashbulb and battery. Again, use of this technique is a matter of trial and error.

As with any experimental technique, extreme caution should be exercised when attempting these recovery techniques. Test these techniques only in large, open areas, with no people or valuable property in the immediate area.

PARAWINGS

Also known as *rogollo wings, parawings* have proven to be quite useful in model rocket recovery. It is similar to a hang-glider in appearance, as shown in Fig. 7-21. A parawing rides inside the rocket's fuselage until ejection, at which time it is deployed by being pushed out of the fuselage and snapped into shape by the air flow.

STREAMER MODIFICATIONS

Streamers have the advantage of permitting very little drift, because they catch little air and permit the rockets to which they are attached to descend very quickly. This is sometimes a disadvantage, though, especially when a streamer is being used with a rather delicate rocket.

There are several streamer modifications that can result in a slower descent and still keep drifting to a minimum. One of the most reliable is to make streamers out of tracing paper, which is very lightweight, yet brittle. The brittleness of tracing paper allows it to be folded in sharp creases and to maintain these folds upon ejection. These extra folds create more drag, thus slowing the descent of the rocket.

Another technique is to use multiple streamers to create extra drag. Again, this technique keeps horizontal drifting to a minimum.

A relatively new development in streamers, about which there is little research data, is the use of what are called *parastreamers*. A parastreamer is essentially a streamer with "pockets" that catch air and create a significant amount of drag. Figure 7-22 illustrates a parastreamer.

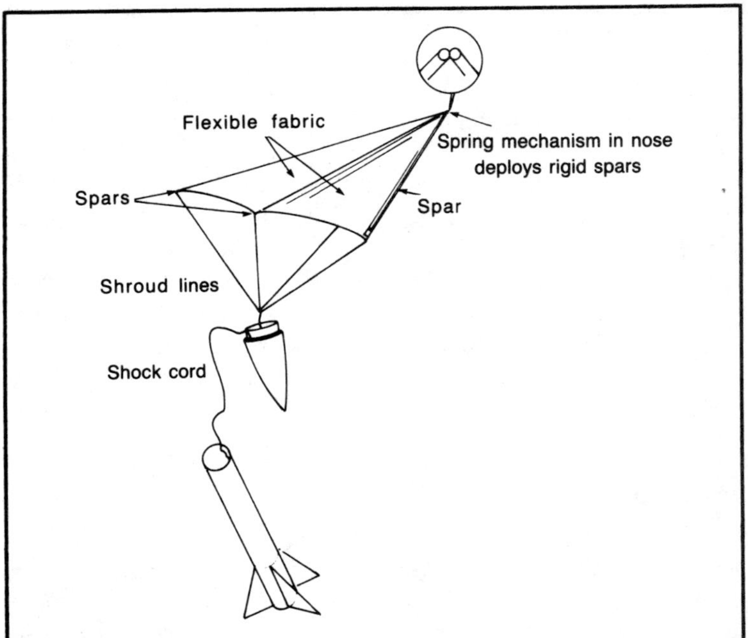

Fig. 7-21. Parawing recovery system.

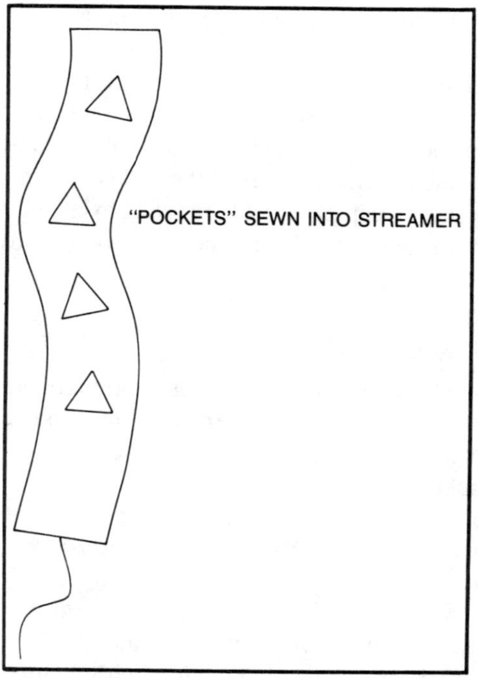

Fig. 7-22. Parastreamer.

UNDERWATER LAUNCHES

An underwater launch is one of the more spectacular things you can do with a model rocket. The rocket to be launched should be carefully prepared by using several coats of flameproof and waterproof glue to fit the parts together, and the entire rocket should be coated with the same glue.

A rocket using featherweight or tumble recovery is recommended, because any opening in the rocket (such as the opening between the base of the nose cone and the top of the fuselage) is likely to cause a leak.

The igniter should be soldered to the wires leading to the launch controller and inserted as far as possible into the engine. After this is done, hot wax or paraffin should be dripped into the engine nozzle and over the igniter until the igniter is completely covered (only the insulated part of the wires leading to the launch controller should be exposed).

This done, a launch rod should be fixed to the bottom of a large container from which the rocket will be launched, either by attaching the launch rod directly to the base of the container (which will probably be a large metal tub) or, better, by attaching the rod to a large block of wood or metal that will set evenly in the bottom of the container.

The rocket should next be placed on the launch rod, resting on the bottom of the container. (Be careful not to disturb the placement of the igniter.) After the rocket is in place, slowly fill the container with water, with the aid of someone to hold the rocket steady.

When these preparations are complete, you are ready to launch. Use the same safety practices and countdown procedure you would with a normal launch.

RADIO CONTROL OF MODEL ROCKETS

Radio control of model rockets during flight is best confined to boost-glider applications. No successful experiments have been conducted using radio control techniques to control the flight of a conventional model rocket.

If you already fly radio controlled (RC) model aircraft, you have a lot of the knowledge needed to fly RC boost-gliders and rocket-gliders. The techniques are basically the same as with aircraft—or, more specifically, as with RC gliders.

Naturally, the lightest possible servos, batteries, and receiver should be used, in order that the model may comply with weight limitations.

Control of the B/G or R/G should be limited to the glide phase of the flight. As with the radio control of conventional model rockets, experiments with control of the powered flight of B/Gs and R/Gs have been largely unsuccessful (not to mention dangerous).

COMMERCIAL AND SCIENTIFIC APPLICATIONS OF MODEL ROCKETRY

The relatively low cost of model rocket technology has made the use of model rockets ideal for some commercial and scientific applications, despite limitations in size, range, and payload capacity.

One obvious commercial use of model rocketry is in movie special effects. Model rockets have been used in the 1960s television series "The Girl From U.N.C.L.E.," the television series "Wonder Woman," and in the film *Meteor* (Fig. 7-23).

Other commercial activities in which model rockets have been used include surveying, line hauling in bridge construction, and aerial photography.

Scientific and research applications for model rockets have included the testing of laser tracking systems, biological monitoring of radiation, and air pollution sampling and measurement. The Motorola Corporation has successfully used model rockets to test radar, as well.

Model rocket prototypes of real rockets and missiles have contributed greatly to the development of rockets and missiles, enabling engineers to "test-fly" their designs before constructing expensive full-size prototypes.

In at least one instance, model rocketry has formed the basis for an entire commercial enterprise. In the early 1970s, Colspan Environmental Systems, Inc., of Boulder, Colorado, offered an entire system, based on model rocket technology, for measuring air pollution over cities. The system (known as LARS, an acronym for Low Altitude Rocket Sonde) consisted of a launch pad, a rocket, a launch control center, and a telemetry ground station.

The rocket itself was made from plastic, about 24 inches in length and 1.24 inches in diameter. Weighing a maximum of 14 ounces, the LARS vehicle carried as its payload a small radio telemetry transmitter, along with a small thermistor that was used

Fig. 7-23. Model rocketry is often used for movie special effects. Courtesy Centuri Engineering.

to measure temperature at various altitudes during the rocket's flight. The receiver for the telemetered information was contained, along with the launch controller, within a suitcase, making the entire system easily portable.

Using the data from air temperature readings at various altitudes (up to a maximum of 4000 feet), meteorologists were able to chart thermal inversions caused by air pollution. Because like all model rockets, the LARS vehicle and its payload were completely recoverable, this system proved to be much more economical than other methods for obtaining the same information, such as from aircraft and balloon flights.

There are, of course, many other research activities that could benefit from similar applications of model rocketry. It remains only for those involved to discover the benefits of the miniature technology of model rocketry.

Chapter 8

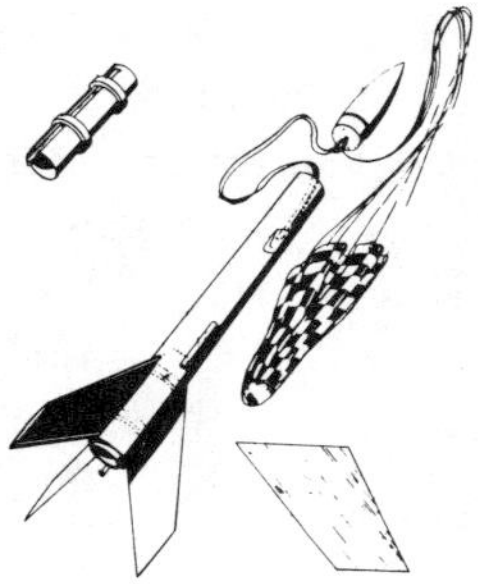

Clubs, Contests, and Other Considerations

Several organizations exist for the benefit of model rocketeers. In the United States, the National Association of Rocketry (NAR) serves the model rocketeer by establishing sporting events and sanctioning national, regional, and local contests. The NAR also certifies engines, conducts a variety of public and educational programs, and publishes a monthly magazine for its members and other interested model rocketeers.

The advantages of being a member of the NAR are many. In addition to keeping its members informed on the latest developments in the hobby through the publication of its monthly journal *the model rocketeer*, and various technical reports, the NAR offers model rocketeers the opportunity to get in contact with other rocketeers and to share model rocketry activities via local clubs, contests, and other activities.

Group discounts are available to NAR members, and low-cost liability insurance is offered through the NAR's affiliation with the Academy of Model Aeronautiques (AMA). Finally, the NAR is affiliated with the Federation Aeronautique Internationale (FAI). This affiliation provides NAR members with the opportunity to set national and international records in events such as altitude, duration, and payload lifting, because the FAI, based in Paris, France, is the international organization responsible for documenting and certifying achievement in real and model aviation and aerospace activities.

In addition to the NAR, you'll find various clubs sponsored by model rocket manufacturers. Although these clubs are very limited in the activities they offer, you may find that joining one or more of them is a good way to stay informed and to take advantage of various "members only" special offers.

Other countries have their own versions of the NAR, most of them allied with the FAI, and participate in the international competition known as the World Spacemodeling Championships, held biannually in various countries. You'll find a listing of the various model rocketry organizations, both in the United States and other countries, along with their addresses, in Appendix B.

LOCAL CLUBS

You may find that a local model rocketry club already exists in your area. Check with local hobby shops to see if such a club exists, or contact the NAR for information on sections in your area (a *section* is a local NAR affiliate).

You may also find that local youth organizations, schools, or service clubs sponsor a model rocketry club. The advantages of belonging to a model rocketry club are similar to those of membership in a national organization. Contact, competition, and information sharing with other rocketeers are the major benefits. Discounts with local hobby shops and model rocket manufacturers are another plus for local clubs. Local clubs can also take advantage of the "demonstration" programs sponsored by some of the manufacturers, which provide clubs with the materials and contacts to carry out public demonstrations of model rocketry.

If there are no model rocketry clubs in your area, but you have several acquaintances who are model rocketeers, you already have the nucleus for a club. If you have no model rocketeer contacts, you can find them best through your local hobby shops. Many hobby shops have bulletin boards, and you can post a notice requesting that any model rocketeers interested in forming a club contact you. Hobby shop owners may also be able to put you in contact with model rocketeers in your area.

Once you have contacted a sufficient number of rocketeers interested in forming a club (usually at least five), you should set a meeting place and time.

Your first meeting will be your organizational meeting, and you should decide on a name for the club, a club constitution or set of rules, and the range of club activities.

Club rules should cover the frequency of meetings, the dues for club members, and the organization of the club in terms of officers, conduct at meetings, etc.

Club activities can include maintaining a launch range, competition, public demonstrations of model rocketry, publishing a newsletter, and building sessions, as well as attending aerospace programs and museums and sponsoring talks by aviation and aerospace engineers, etc. The range of activities will depend upon the interests of the club membership and the willingness and ability of various members to plan and work on activities.

COMPETITION

Competition in model rocketry can take many forms, from attempting to achieve the highest altitude with a payload to trying to keep a B/G aloft for the longest period. The more popular contest events are outlined below.

In each case, the event can be divided into different divisions, based on engine size. (For example, you may wish to hold one competition for model rockets powered by A engines, another for B engines, and so forth. Multiple stage rockets should also be in a class by themselves.)

Altitude Competition. Competition to reach the highest, or maximum, altitude is probably the most popular type of contest. Each rocket should be launched separately and tracked using the tracking techniques outlined in Chapter 5. An optional payload (consisting of a 1-ounce lead weight or sandbag) may be used, if the rockets have payload compartments.

Parachute Duration. In this type of competition, the object is to achieve the longest time in the air, beginning with the first motion of the rocket on the launcher. The time is measured from that point until the rocket either touches down or drifts out of sight.

The strategy here is, of course, to build a model rocket as lightweight as possible and with a parachute that will provide the most support with the least amount of weight. A high initial altitude is also helpful, so the model rocket should be as low-drag as possible.

Streamer Duration. This competition is similar to parachute duration, the major difference being the use of a streamer instead of a parachute. Long, wide, lightweight streamers are the most efficient. A rocket designed to separate at mid-fuselage is also helpful (the two parts remaining joined by a shock cord, of course), because this creates more drag and slows the rocket's descent.

Glider Duration. This is a contest for boost-gliders or rocket-gliders. As with the other events, rockets should be classed by engine size. The glider is timed from first motion on the launch pad until it either touches down or drifts out of sight.

Payload Lifting. The standard payload for payload lifting competition is a 1-ounce lead weight or 1-ounce sandbag. The object of this competition is to attain the maximum altitude carrying the weight.

An alternative payload is an egg—fresh, uncooked. In addition to a high altitude, the goal in carrying an egg—also known as *egg-lofting*—is to recover the egg unbroken. A point system can be established to score this contest, with so many points per foot of altitude, and so many points for recovering the egg intact.

In each type of payload competition, the strategy is to build as efficient a rocket as possible. In the egg-loft, there is an additional challenge in the need to design a payload compartment that can protect the egg from the stresses of flight and recovery.

Spot Landing. The purpose of this competition is to land as close as possible to a set landing point, marked by a flag or other device.

The landing spot may be selected by arbitrary choice or by launching a small tumble- or featherweight-recovery rocket and designating its landing spot as the landing goal.

The spot landing competition is usually open to any type of rocket, powered by any engine or combination of engines. There are no classes.

Scale. Scale modeling is perhaps the most difficult and, at the same time, the most rewarding model rocket competition. The primary purpose of scale modeling competition is to build a model rocket that duplicates as closely as possible the appearance of a real rocket or missile. The prototype should be a working "round"—not a display rocket or missile.

Points are awarded for scale data, overall appearance, detailing, and flight performance, generally. Scale data can include such items as color or black-and-white photographs, original plans or drawings, and published descriptions. Overall appearance takes into account the model's finish, the craftsmanship that went into the model, and its faithfulness in dimensions to the original. Detailing also takes into account the model's craftsmanship, as well as any working parts or small details such as attitude jets, vanes, etc.

Flight performance can be judged in several areas. Scale competition actually breaks down into several events, based on which aspect of flight performance is being judged. There may be a scale altitude event, a scale payload event, or even a scale duration event. (These should be classed according to engine size, of course.)

Research and Development. This competition can take place in any of a number of categories and involves new techniques in flying and/or building model rockets, new designs in recovery or guidance systems, new tracking techniques or devices, or developments in any other area of model rocket flight, construction, or performance.

R&D, as it is commonly known, is judged on a number of factors, including presentation of theoretical data through displays or discussion, performance of the new design or technique, and effectiveness of the design or technique, in addition to the validity of any theories.

Static Spacemodeling. A relatively new competition has surfaced in recent years involving the construction of nonflying model rockets. This is *static spacemodeling.* Static spacemodeling generally falls into one of three catagories: kit building, prototype modeling, or open design. *Kit building* requires the use of a commercially available kit and may be subdivided into modified or "stock" categories. *Modified kit building* permits the use of parts that are not included with the kit, while stock requires that only the parts included with the kit be used. This is mainly a craftsmanship event.

Prototype modeling is the design and construction of an aerospace vehicle or spacecraft model that is based on a real prototype. Many of the considerations involved in the judging of a scale model apply in judging a model in this event.

Open design, sometimes known as "freestyle," permits the design and building of any type of aerospace or space vehicle, including modified versions of existing vehicles, models of proposed spacecraft, or designs of the modeler's own creation. Here, the judging parameters include originality and craftsmanship.

These are the major events, used by most clubs in competition. You may modify these to suit your needs. If you are participating in NAR sanctioned events, you will find that there are very specific rules for these and other events, set forth in the United States Model Rocket Sporting Code, a booklet of rules compiled and published by the NAR (provided to all members).

If you wish to participate in NAR sanctioned competition, and thus compete for National point standings, you must be an NAR member. (Special one-day memberships are available for competitions.) The NAR sanctions local, regional, and national meets, and points are awarded in a cumulative national standing system for winning in various events. National standings are also recorded for local NAR sections.

The NAR sponsors one national meet per year, known as NARAM (National Association of Rocketry Annual Meet). In addition to this, NAR sponsored teams compete in international competition at the World Spacemodeling Championships. These teams are selected by a biannual competition preceding the WSMC. These events are open to only full-time NAR members.

Competition is perhaps the ultimate activity in model rocketry, requiring competitors to test their design, construction, and flying techniques to the limit. It is also highly rewarding in itself.

INSURANCE

As mentioned previously, NAR members can obtain liability insurance for their model rocket activities at a very low cost. Although model rocketry is, statistically, safer than baseball, and there has never been a claim against this insurance, it is still advisable to obtain insurance as a NAR member.

MODEL ROCKETRY AND EDUCATION

Model rocketry offers many opportunities in educational applications and has been used from the grade school level through high school and college.

In addition to being an aid in the study of rocketry, model rocketry can easily illustrate the Laws of Motion for physics or science classes. Working with the applied mathematics of model rocketry is a useful enhancement to the study of various levels of mathematics.

Model rocket manufacturers offer several books suggesting applications of model rockets in the areas of math, science, shop, and even literature.

College-level educational applications of model rocketry have included programs in the basics of rocketry at institutions such as the United States Air Force Academy and the University of California at Los Angeles' summer program. Aimed at providing participants with a basic knowledge of the aerodynamics and physics principles involved in model and real rocketry, these programs have been highly effective.

Grade school and high school programs in model rocketry generally function as supplements to other areas of study. For example, a high school physics class may build and fly model rockets as an aid in the study of the Laws of Motion, and a shop class may approach model rocketry as a craft. Model rocketry programs in elementary and higher level education may also serve as an aid in studying space exploration and real rocketry.

Some school programs, and many youth group programs, such as 4-H and various summer camps, offer model rocketry as an end in itself. The various activities involved in building and flying model rockets provide an education in themselves.

In almost any educational application, model rocketry can be counted on for results. The "hands-on" nature of the study of a topic through model rocketry invites participation and maintains interest.

For those interested in using model rocketry in support of science studies, or as a topic of study in itself, it is suggested that the history of rocketry and spaceflight, along with the basic principles of flight, be employed in the program of study. The level of study is, of course, dependent upon the age group of the students. Elementary students can be exposed to the basic principles of flight by example, while high school and college programs might become more involved with the physics and explanatory mathematics of flight and rocketry.

The following syllabus is one approach to presenting model rocketry as a part of a unified study of aerospace science.

I. **Background**
   A. History of flight.
   B. History of rocketry.
   C. Basic principles of rocketry.
      1. First Law of Motion.
      2. Gravity.
   D. Basic principles of flight.
      1. Lift and drag.

II. **Rocketry**
   A. Parts of a rocket.
   B. Rocket performance.
      1. Thrust.
      2. Drag.
      3. Aerodynamic principles affecting a rocket in flight.

C. Model Rocketry.
   1. Definition of model rocketry.
   2. Presentation of model rocketry technology. (Engines, components, etc.)
   3. Terminology.
   4. Design and construction of model rockets.
   5. Class constructs model rockets.
   6. Class flies model rockets.
   7. Discussion of flight data.

III. **History of Spaceflight.**

A. Philosophy of spaceflight.
   1. Early myths and legends.
   2. Nineteenth century viewpoints (Jules Verne and other fictional treatments).
   3. Twentieth century viewpoints (science fiction).

B. Spaceflight pioneers.
   1. Robert Goddard, Tsiolkovsky, Werner von Braun, et al.

C. Manned spaceflight chronology.

D. Study of manned spaceflight techniques.
   1. Early Russian and U.S. programs, utilizing space "capsules."
   2. Space stations.
      a. Skylab.
      b. *Salyut* (Russian space station).
      c. Spacelab (European Space Agency/NASA joint venture in space research).
   3. New generation of space vehicles.
      a. Space shuttle.
   4. The future of space travel.

Again, the amount of detail and degree of difficulty of study in a program such as this is dictated entirely by the educational level of the students involved. In any program, model rocketry will provide a graphic presentation of the principles of rocketry, as well as an overall enhancement to the program and an inducement to further study.

With all age groups, the model rocketry section of the program may be expanded to include various competitions, using the contests outlined earlier in this chapter. If you wish to conduct altitude contests, the class may also become involved in constructing altitude tracking equipment, using the plans in this book.

Junior high and high school students may find the design and construction of ground system equipment of interest. (Plans for launch controllers and pads are included in this book.)

For high school and college classes, advanced model rocketry projects can be quite rewarding. Students may be asked to create their own research projects, or they can be assigned projects involving aerial photography and photo interpretation, biological experiments (using insects as payloads), computer simulation, or atmospheric research.

Model rocketry can also be used in support of disciplines other than science. English students, for example, can be assigned to produce newspaper style coverage of model rocket research and flights, or, perhaps, more comprehensive reports of model rocketry programs, as well as reports on the history of rocketry and spaceflight.

Industrial arts programs can be supplemented by model rocketry activities, including the design and construction of ground system equipment communications and tracking systems, and launch range layout and support equipment.

Both Estes Industries and Centuri Engineering Company have produced general and specialized manuals for educators.

Studies involving model rocketry can be used as a springboard to other related activities. A logical extension of model rocketry and/or rocketry, spaceflight, and aerospace studies is the history and politics surrounding the development of rocketry. Field trips to attractions such as aerospace museums, military bases, and operational rocket ranges or tracking centers can be a part of this study.

Aerospace manufacturers can be another resource for the study of rocketry and spaceflight. Many will provide speakers as well as educational materials.

Educational activities using model rocketry are varied and nearly limitless and lead to a number of additional areas of study. Using model rocketry as a motivational focal point, it is possible to conduct a successful program with nearly any age group.

Chapter 9

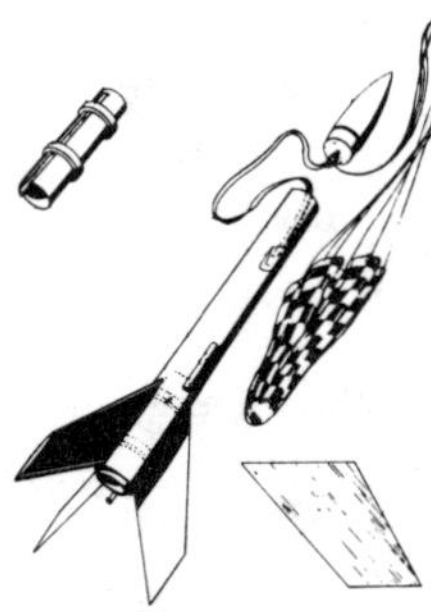

Hints and Tips

As with any endeavor, there is almost always a better, cheaper, or faster way to do something in model rocketry. The information presented in this chapter, collected at random, is intended to help save you time and money and avoid problems in your model rocketry projects. With a little experience, you'll come up with many more similar techniques on your own!

- Don't buy kits exclusively. Although some model rocket kits, such as Estes Industries' Astrocam, contain parts that are not otherwise available, most model rocket kits are composed of body tubes, nose cones, and fin stock that the manufacturers sell separately. In not buying a kit, you normally miss out on the special decals, the instruction sheets, and the occasional plastic part, and that is all. As a less expensive alternative, try building your own designs occasionally, or designs that are close to the kits you'd like to build.
- Use "pop-n-go" engine mounts, like the one illustrated here (Fig. 9-1). This will let you use one engine mount for several rockets with the same diameter body tube, instead of making an engine mount for each rocket.
- Use basswood instead of balsa wood in your own designs. Basswood costs about the same as balsa, but it is several times stronger, although heavier. The benefit here will be the fact that the fins of your model will be almost indestruc-

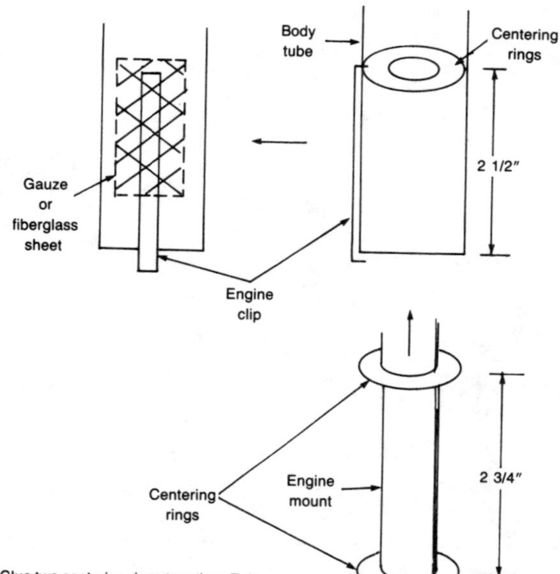

Glue two centering rings together. Enlarge center hole so the top of the engine mount can pass through. Insert one end of an engine clip into the body tube, 2 1/2" from bottom. Glue the rings inside the tube, against the clip. Cover about one-half of the exterior engine clip with white or yellow glue, then apply a piece of gauze to the clip.

Fig. 9-1. Pop 'n' Go engine.

tible when compared to balsa fins. This will save you time and money in repairs.

- Make your own parachutes from garbage bags (Fig. 9-2). Cut the parachute from the bag in the same hexagonal or octagonal shapes of premanufactured parachutes. Use nylon fishing line for shroud lines and attach the shroud lines to the parachute with electrical tape. Use a snap-swivel of the type used to attach hook lines to fishing lines to connect the parachute to the nose cone of your rocket.
- Use snap-swivels on your streamers and parachutes so that you can use them with more rockets. This will mean fewer parachutes to make, and you will have the capability to quickly replace a parachute or streamer should one be destroyed during a flight (Fig. 9-3).
- If one of your rockets "bites the dust," salvage as many parts as possible. Fins, engine mounts, engine clips, etc., may all be reused.

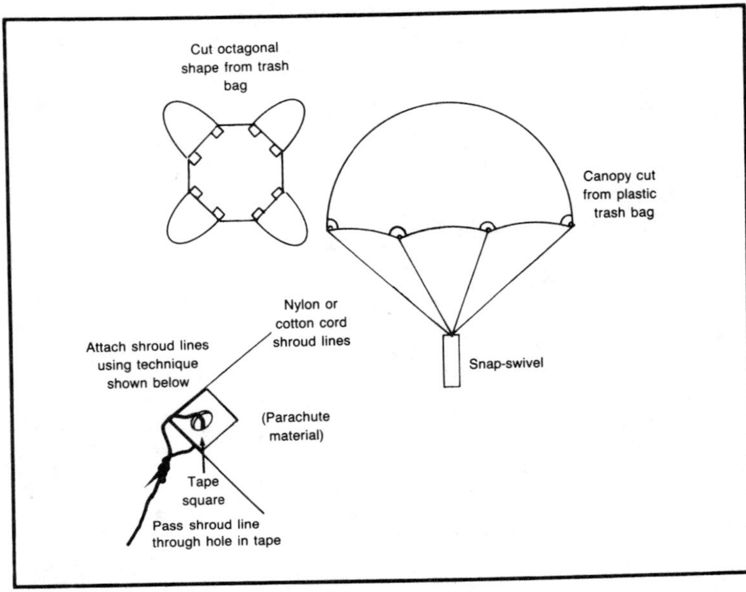

Fig. 9-2. Homemade parachute.

- Repair kits whenever possible. The accompanying Figs. 9-4 and 9-5 will give you some ideas on how to repair damaged model rockets.
- If you need weight for the nose cone of a model rocket and you don't have clay, make your own nose weight instead of buying it. White glue, mixed with sand or other material,

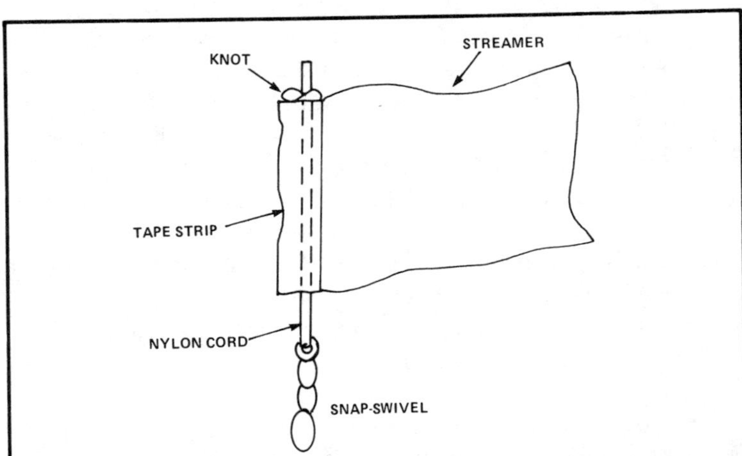

Fig. 9-3. Reusable streamer.

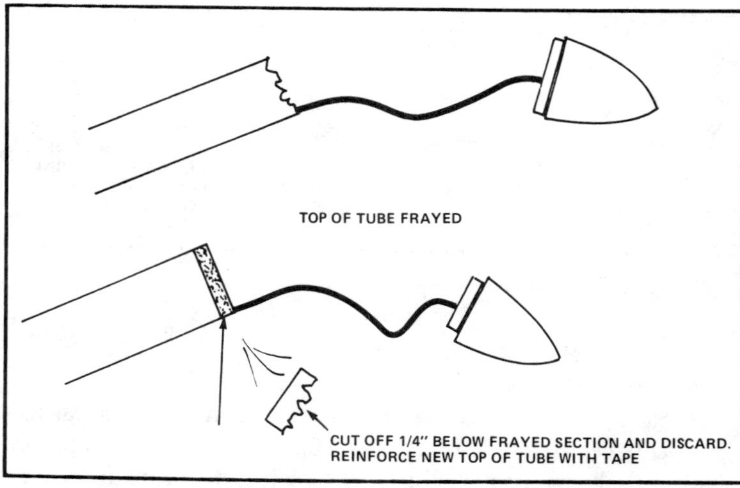

Fig. 9-4. Repairing damaged body tube.

can be poured into a nose cone and will stay there once it hardens.

- Instead of buying parachute powder (dusting a parachute with powder helps it open fully in cold weather), use less expensive baby powder or corn starch.
- Make shock cords from the threaded elastic material sold in sewing departments of stores. It is a cheaper and stronger material than the standard rubber band-type shock cords sold

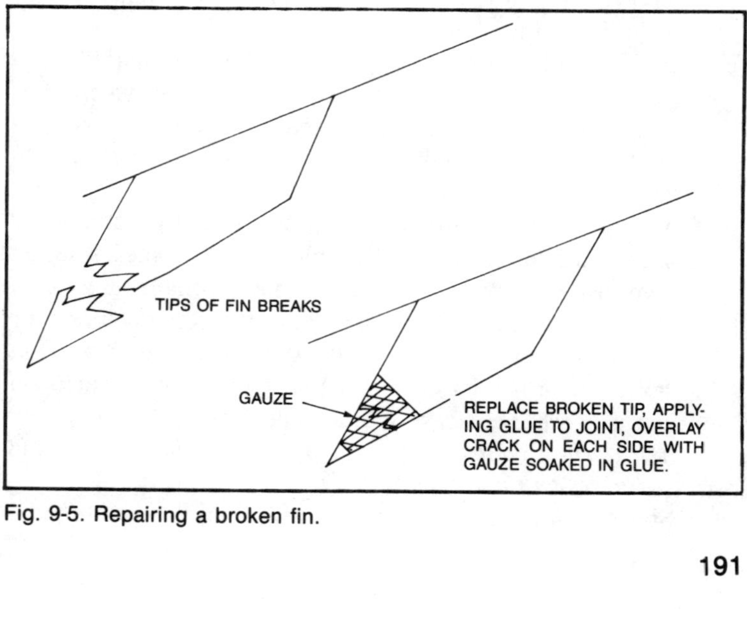

Fig. 9-5. Repairing a broken fin.

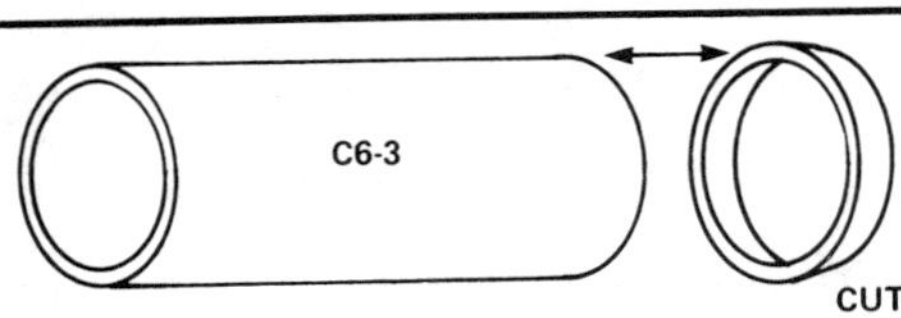

Fig. 9-6. Using old engine casing for thrust rings.

by model rocket manufacturers.

- Use plastic soda straws for launch lugs.
- Save your expended engine casings. The mini-engine casings can be used to make launch lugs for large, high-power models and can also be used to cap launch rods for safety when the rods are not being used. Sections of larger casings can be used as thrust rings, as in Fig. 9-6, or as engine mounts.
- Launch rods can be made from 3-foot- × -3/8-inch diameter welding rod or piano wire.
- In a pinch, paper clips can be used in place of micro-clips to connect launch controller wires to igniters, as in Fig. 9-7.
- If you're having trouble getting regular spray or brush enamel to stick to plastic parts, try spraying a coat of automotive primer on the plastic parts first. It is easy to apply, dries fast, and makes an excellent foundation for painting the final color.
- If you're a bargain hunter, check hobby shops and department stores for "buys" on model rockets and accessories in the fall and right after Christmas. Those stores that normally offer seasonal discounts will discount model rockets at these times.
- You can make your own flameproof recovery wadding by mixing a solution of boric acid and water to make a solution two to three times as strong as that recommended as an eyewash on the bottle. Soak toilet tissue in this solution for at least one minute. After soaking the paper, set it aside in a pie pan or similar container to dry for about a week. Do not squeeze the excess solution out of the paper. *Caution: Boric acid is poisonous; be sure to follow the instructions and warnings on the bottle carefully.*
- Spend some time surveying the model airplane and model

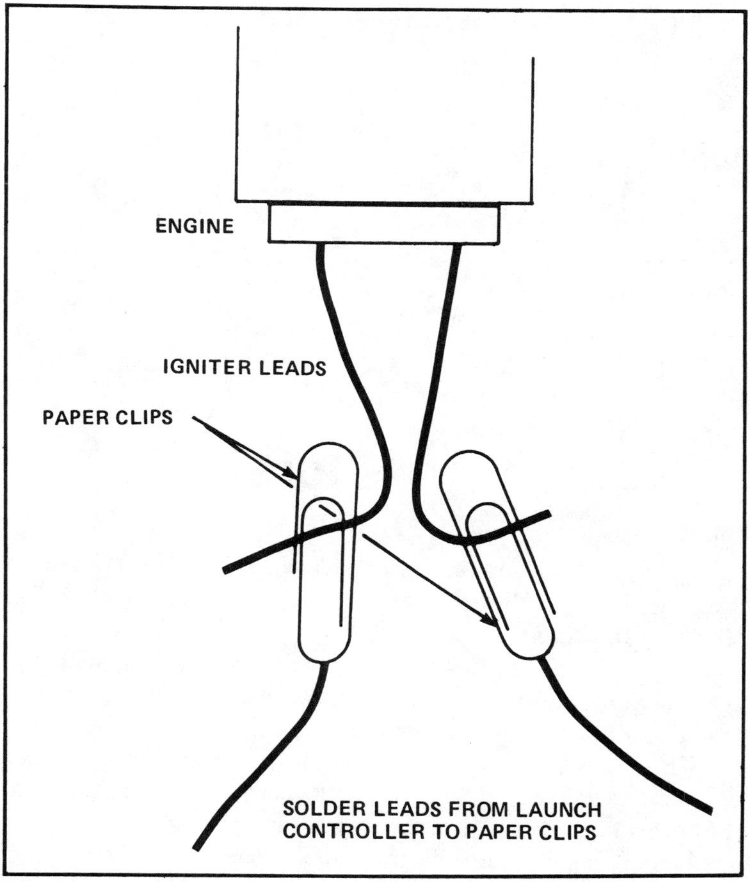

Fig. 9-7. Use paper clips instead of micro-clips in an emergency.

railroading sections of your hobby shop. You may find many items and accessories in these areas that are excellent for use with model rockets. They may be cheaper. You may also find useful items in the notions department of a store, and even among the plastic containers discarded with your trash!

Chapter 10

The Future

Model rocketry has, for the most part, been a growing hobby, although it has had its ups and downs like anything else. The period after the slowdown of the American space effort in the early 1970s was a particularly slow one for the growth of the hobby.

The increasing availability of leisure time, along with the renewal of interest in space due to space shuttle activities, should result in a new spurt of growth for the hobby.

This growth will not be restricted to an increase in the number of hobbyists and the volume of activity. The range of activities and the technology of model rocketry will also expand.

Model rocketry's technological development has always been closely tied to the activities of hobbyists, as has that of model aircraft. Manufacturers, such as Estes Industries, have consistently promoted development on the part of the end users of their products. For example, many of Estes' products first existed as entries in the "Design of the Month" contest, which has been an Estes sponsored activity for years.

Those who work with the model rocket manufacturers are, for the most part, enthusiastic supporters of and participants in the hobby, so it might be said that here, too, there is input into the growth of model rocketry from hobbyists. Also, some of the smaller, specialized model rocket manufacturers and suppliers were started by dedicated hobbyists and remain in existence today because of their dedication. With this element in place, an increase in the number of hobbyists can only mean advancements in model rocketry technology.

What forms might these advancements take? Already there is quite a bit of research and development being carried out by hobbyists in the area of aerial photography. Eventually, there should be an autowind camera available for use with model rockets, one which is capable of taking more than one photo per flight. There may even be a replacement for the defunct Cineroc, the aerial motion photography camera. Hobbyists are working in both of these areas, and when the time is right, such cameras will appear on the market.

Hobbyists are also working extensively with electronic gadgetry, developing such things as computerized launch controllers, radio control devices, miniaturized timers for parachute delay and B/G applications, and timing and tracking devices. Products such as these, too, will appear on the market when development is complete and public demand is shown. Some already are.

Engine development, while slow, should pick up in the future, in response to market demands for a variety of high- and low-power engines for payload and other applications.

Kits, of course, will always be available in new and endless variety. Manufacturers, as a part of the marketing process, are constantly withdrawing old kits and replacing them with new kits, in response to sales, projected interests, and developments in real space and aerospace vehicles. The acceleration of space activities by the United States, Europe, and the Soviet Union will surely bring a plethora of new scale and sport models, inspired by real spacecraft and aircraft.

Model rocket component technology should also advance, as manufacturers seek more reliable materials to please customers and more cost-effective processes and materials for their own benefit. Plastics as a substitute for cardboard and balsa hold great promise, and advances in paint and glue formulas should also materialize.

As the technology of model rocketry improves, we should see advancements in competitions. Specifically, new records in altitude, duration, and other areas should result from new technology. New competition events may also result.

Other areas that have potential application in model rocketry include computers (onboard computers for data gathering and storage, computer analysis, and—possibly—control of model rocket flight performance), television (onboard cameras and/or recorders), and large-scale integration in electronics.

One aspect of model rocketry that will doubtless remain unchanged, though, is the fact that model rocketry is fun!

Appendix A

Sources of Model Rockets and Supplies

Celestial Enterprises 4412 Crabapple Street, Forth Worth, TX 76137	*Kits, parts. Write for product list.*
Crown Rocket Technology (Formerly SSRS) Box 341 Mountlake Terrace, WA 98043	*Specialized manufacturer of high-power kits and engines. Catalog $1.50.*
Estes Industries, Inc. Box 227 Penrose, CO 81240	*Full line model rocket manufacturer: kits, engines, parts, publications, and supplies. Catalog 75¢.*
Flight Systems, Inc. (FSI) 9300 East 68th Street Raytown, MO 64133	*Specialized manufacturer of high-power kits and engines. Catalog $1.50.*
High Power Research P.O. Box 73 Cuba, OH 45114	*Supplier of high-power kits, engines, and literature.*

Pinecap Associates
P.O. Box 228
Derby, KS 67037

Supplies mainly high-power kits and engines at a discount. Catalog 75¢.

Appendix B

Model Rocketry Organizations

UNITED STATES

Estes Aerospace Club (EAC)
Estes Industries, Inc.
Box 227
Penrose, CA 81240

National Association of Rocketry (NAR)
182 Madison Drive
Elizabeth, PA 15037
(Note: the National Association of Rocketry provides a listing of local clubs in the United States to its members.)

Chrix Tavares
Model Rocketeer
8468 Falling Leaf Road
Springfield, VA 22153

INTERNATIONAL

Australia

Ken McNab
18 Daniel Street
Glen Waverly, 3150
Victoria, Australia

Bulgaria

Angel Jankov
Bulgarian Federation of Rocketmodelers
1000 Sofia
Boulevard Christo Botev
48 Bulgaria

Canada

Canadian Association on Rocketry (CAR)
Suite 302
151 Slater Street
Ottawa, Canada
K1P 5H3

Peter Cook
35 Warrender Avenue
Islington, Ontario
Canada M9B 5Z5

Czechoslovakia

Otakar Saffek
Hajkova 14
13000 Praha
3 Czechoslovakia

England

Martin Dilly
22 Links Way
West Wickham
Kent, England

Poland

Edward Kurowski
Wrzeciono 6-11
01-651 Warszawa
Poland

Spain

Angel Infante Moratilla
Federation Nationale de los Deportes Aereos
Subcommittee de Modelismo Espacial
c/o Ferraz No. 16
Madrid, Spain

West Germany

Wolfgang Carstens
Wolfsberg
D-2359 Hasenmoor
West Germany

Yugoslavia

Srdjan D. Pelagic
Sekspirova 36/I
21000 Novi Sad
Yugoslavia

Glossary

acceleration—A change in the velocity of an object. *Acceleration* is commonly a positive change in velocity, or an increase in speed, while *deceleration* indicates a negative change in velocity, or slowing down.

accelerometer—A device that measures acceleration.

airfoil—A surface, such as a wing or fin, which is designed to provide lift, control, or a stabilizing effect.

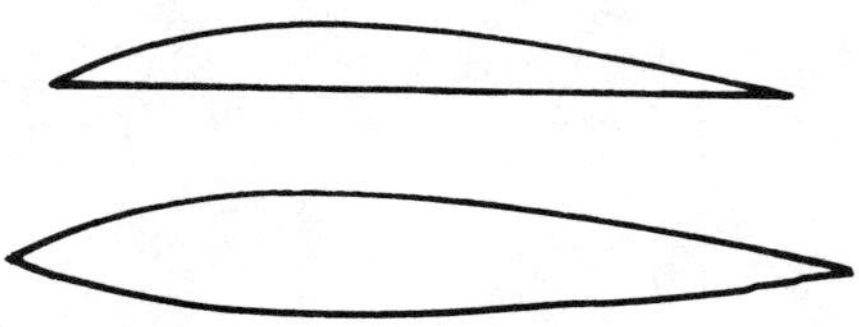

Airfoil cross sections.

amateur rocketry—An activity that involves the mixing of solid- or liquid-fuel, and the fabrication of rockets from metal, amateur rocketry is largely carried out by those who have been trained in such specialized fields.

apogee—The highest point in a rocket's flight.

average impulse—The amount of power available from a model rocket engine at any given point in time.

balsa wood—An extremely lightweight wood, with a high strength-to-weight ratio that makes it ideal for model building.

blast deflector—A shield, plate, or conduit made of metal, which deflects a model rocket's exhaust away from the ground and the rocket while protecting the launch pad.

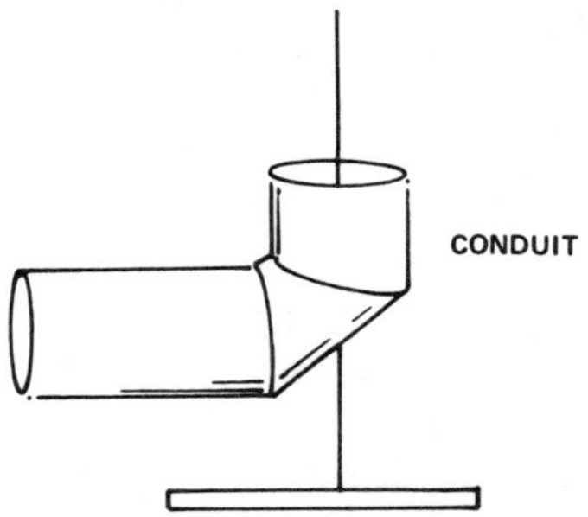

Blast deflectors.

body tube—A cardboard or plastic cylinder that comprises the fuselage or air frame of a model rocket.

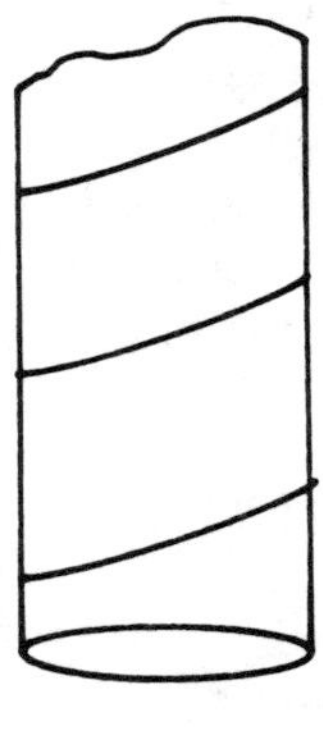

Body tube.

boost-glider—Also known as a B/G, a boost-glider is a model rocket which, through a change in its weight distribution or form, becomes a glider at apogee. A boost-glider is characterized by the fact that it ejects its engine after burnout, unlike a rocket glider *(q.v.)*.

burnout—The point at which a model rocket engine's propellant charge is exhausted, and the engine no longer produces thrust.

center of gravity—The point about which a body acted upon by gravity is in equilibrium in all positions. The point of balance of a model rocket. Symbol:↑

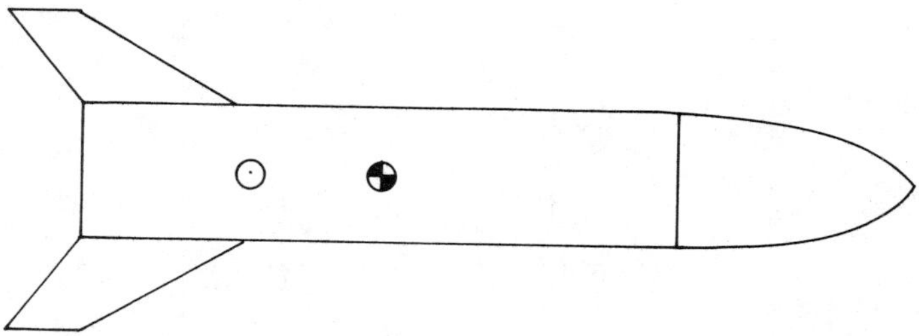

CP and CG.

center of pressure—The "balancing point" for all aerodynamic forces acting on a model rocket. Symbol: ↑

clustering—A technique involving firing two, three, or four model rocket engines simultaneously in a model rocket.

delay or delay element—A slow-burning, smokey charge that is ignited in a model rocket engine after burnout. This provides an aid to tracking a model rocket and provides a time delay to permit the rocket to coast to apogee.

dethermalizer—A device—usually a small fuse—that is activated at liftoff and provides a method of destroying a glider's lift after a predetermined time so that the glider may be safely recovered.

drag—Aerodynamic force on a model rocket, which causes it to slow down, or, the resistance of a model rocket to air flow.

ejection—The act of ejecting or forcing an object from a container, such as ejecting a parachute from a body tube.

ejection charge—A charge in a model rocket engine that is activated by the delay element to produce a miniature explosion. The hot gases generated by this explosion expand in the body tube and force the parachute out, in the case of parachute recovery model rockets.

engine—A miniature solid fuel rocket motor with nonmetallic components.

FAI—*Federation Aeronatique Internationale*, the Paris, France, based organization that documents world records in real and model aviation and aerospace activities.

fin—In model rocketry, an airfoil attached to the fuselage of a model rocket that helps guide the rocket in flight.

fuselage—The body, or main component, of a model rocket, also known as the air frame.

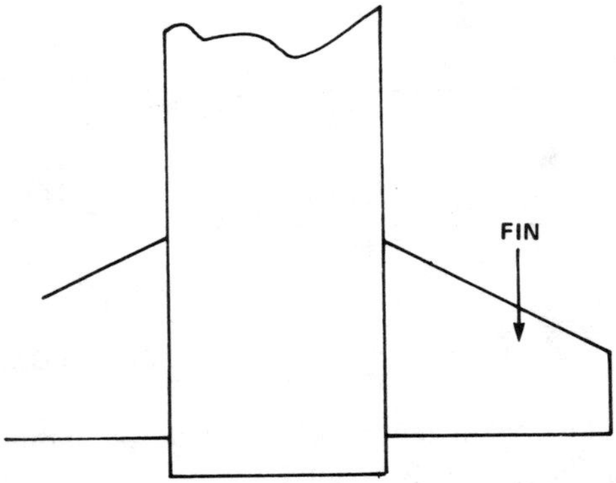

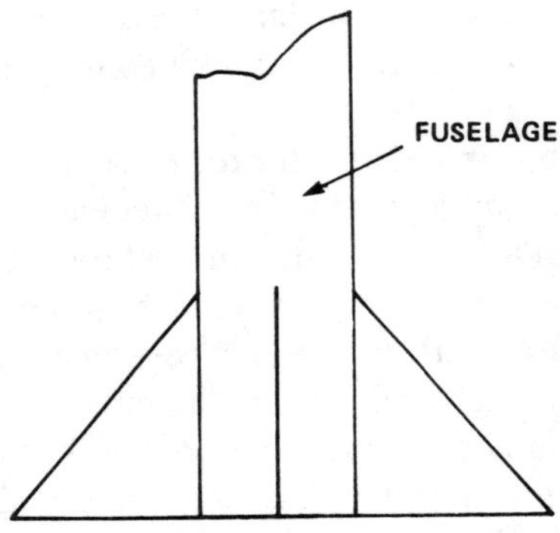

Fuselage.

G—Abbreviation for "gravity." Also a measure of force of acceleration (G-force).

grain—In a sheet of balsa, the direction of the fibers or fiberous particles in the wood and the pattern made by these fibers.

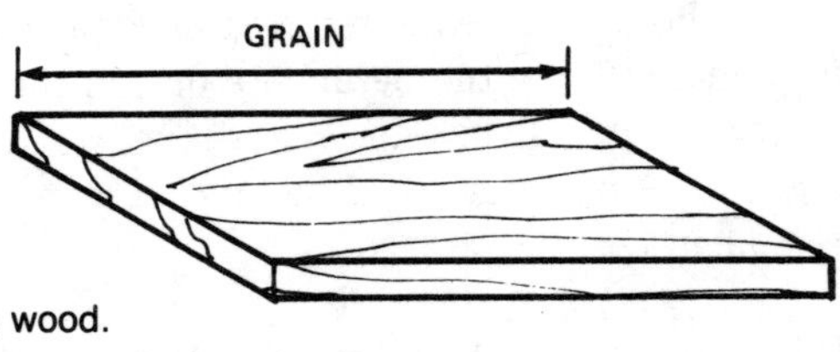

Grain in a piece of balsa wood.

GSE—Ground support equipment, consisting of launch controller, launch pad, and associated components.

high-power rocketry—In general, model rocketry involving the use of any engine or combination of engines higher than a D, or with a total impulse of more than 20 newton-seconds.

igniter—An electrical device consisting of high-resistance wire and pyrotechnic material that is used to ignite a model rocket engine.

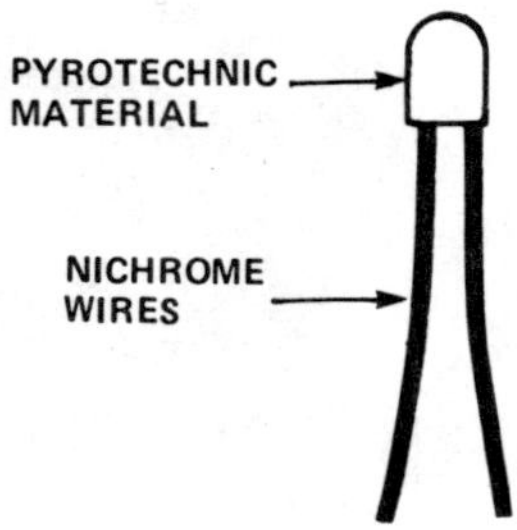

Igniter.

impulse—Force producing motion.

launch—The liftoff of a model rocket after engine ignition.

launch controller—An electrical device, which is used to switch current to a model rocket engine igniter. Launch controllers normally have several safety features, such as a spring-loaded actuating switch, a continuity indicator lamp, and a safety interlock key.

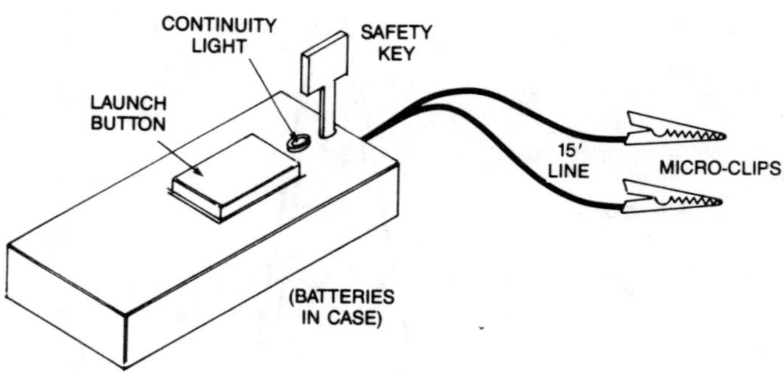

Design of a typical launch controller.

launch lug—A small tube attached to the fuselage of a model rocket which, when placed over the launch rod, guides the rocket during the first few feet of its flight.

launch rod—A length of wire, 1/8- to 1/4-inch thick, which is used to guide a model rocket during the first few feet of flight.

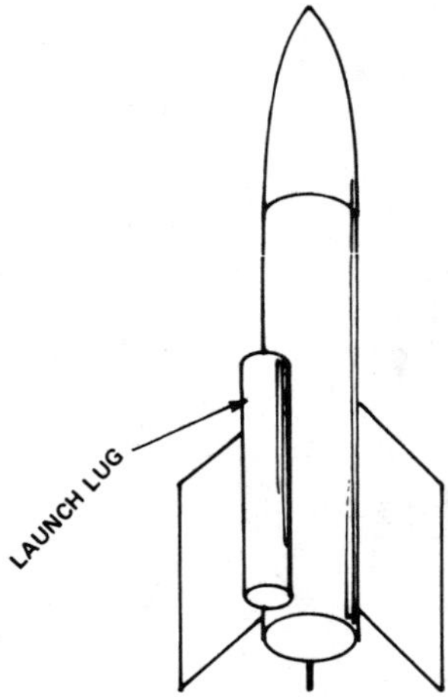

Launch lug.

mass ratio—The ratio of the mass of a rocket at liftoff to its mass after its fuel has been burned. A mass ratio of 2:1, for example, indicates that the rocket's fuel comprises half of its total weight.

micro-clips—Small alligator clips with flattened jaws that are used to connect the wires from a launch controller to igniter leads.

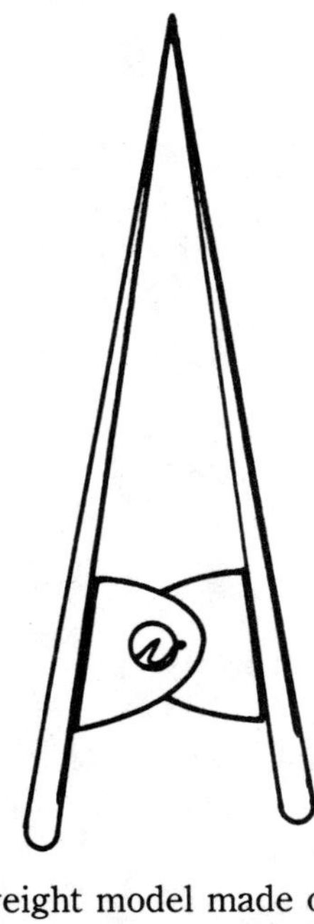

Micro-clips.

model rocket—A lightweight model made of frangible materials and powered by a miniature solid-fuel rocket motor.

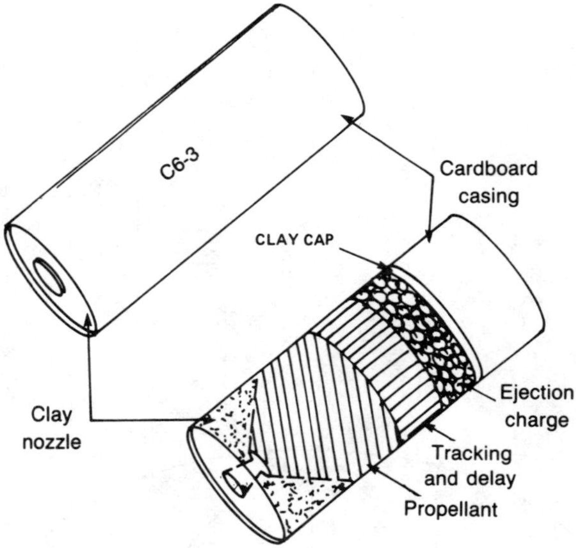

Model rocket engine.

multistage rocket—A rocket with more than one stage.

mylar—A strong, extremely lightweight plastic used for streamers and parachutes.

NAR—The National Association of Rocketry.

newton—A metric unit of force, which must replace the pound as a unit of force in most applications. (4.45 newtons of force equals 1 pound of force.) Specifically, one newton of force (abbreviated n) will induce an acceleration of 1 meter per second^2 in a weight of 1 kilogram.

newton-second—The metric unit of impulse, a counterpart of pound-second. Newton-second denotes one newton of force applied for one second.

nichrome—A type of high-resistance wire that becomes very hot when an electrical current passes through it. Nichrome wire is used in model rocket igniters.

nose cone—The forward section of a model rocket, roughly cone-shaped, sometimes pointed and sometimes rounded.

nozzle—The opening in the base of a model rocket engine, usually made of clay, which directs the flow of exhaust gasses.

parachute—An umbrella-shaped device, usually made of lightweight fabric, used to slow the descent of a body through air.

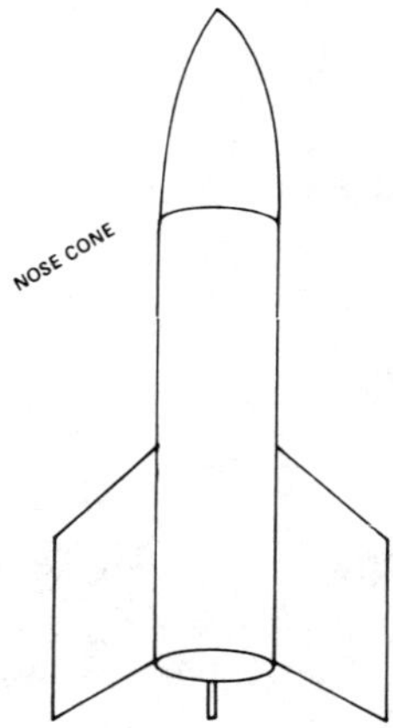

Nose cone.

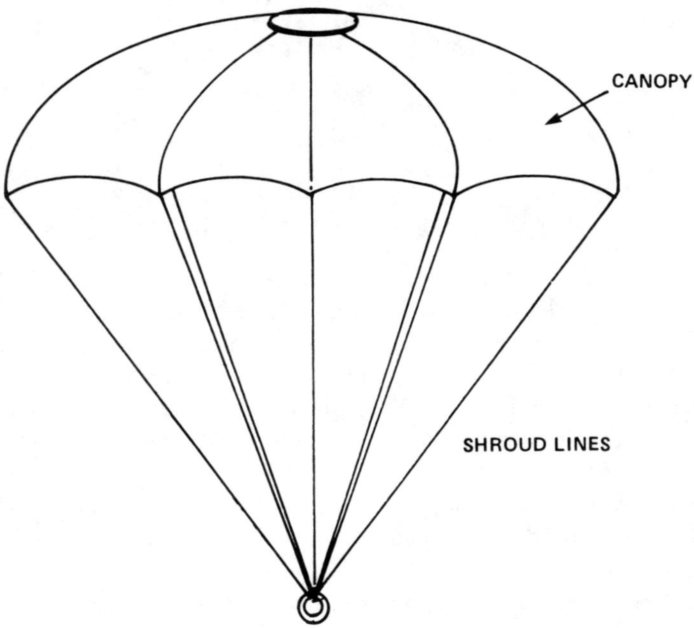

Parachute.

pound-second—The English unit of total impulse, now replaced in general usage by the newton-second (*q.v.*).

payload—The cargo of a rocket.

propellant—The source of motive energy in a model rocket engine.

rocket glider—Also known as an R/G, a rocket glider is a model

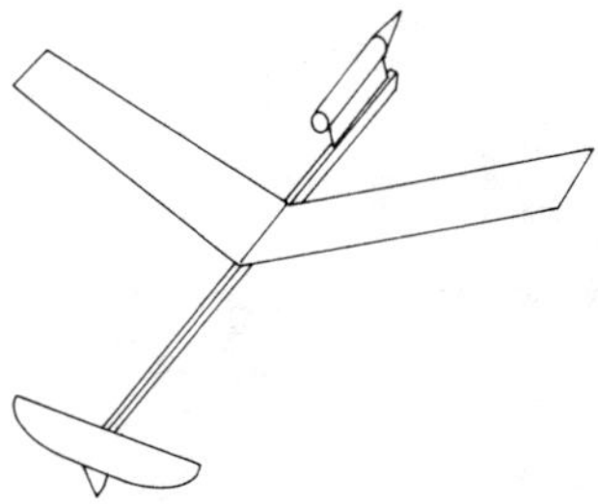

Rocket-glider (R/G).

rocket which, through a change in its weight distribution or form, becomes a glider at apogee. A rocket glider is characterized by the fact that it does not eject its engine at burnout, unlike a boost glider (*q.v.*).

scale model—An extremely detailed model that attempts to duplicate in an exact scale relationship the dimensional relationship and features of a working "real" prototype.

shock cord—An elastic cord connecting a model rocket to its recovery device. The shock cord absorbs the force of ejection, protecting both the rocket and the recovery system and preventing separation.

shroud lines—The lines connecting a parachute to the device it is slowing in descent, or to the shock cord.

specific impulse—The thrust of a rocket, measured as the value per unit rate of a propellant, given in pounds weight per second per pound.

streamer—A strip of mylar, paper, or other material that serves as a recovery device in lieu of a parachute when a fast descent is desired.

thrust—The force that a model rocket engine experts.

total impulse—The total available thrust or impulse of a model rocket engine.

wadding—Flameproof paper or cotton, or heat-resistant plastic, which is used to protect the recovery system of a model rocket from the hot exhaust gases of a model rocket engine's ejection charge.

wind tunnel—A large chamber through which air is forced at selective velocities. A model of an aircraft or aerospace vehicle can be placed in a wind tunnel, and the aerodynamic forces acting on a real vehicle can be simulated through the airflow.

Index

S

T

U

V

W

Y